UNDERSTANDING SCIENCE

YEAR 11
SECOND EDITION

ROY MATHEWS

NELSON
CENGAGE Learning™

Australia • Brazil • Japan • Korea • Mexico • Singapore • Spain • United Kingdom • United States

Understanding Science Year 11 Level 1
2nd Edition
Roy Matthews

Cover design: Cheryl Rowe
Text design: Cheryl Rowe
Illustrations: Cheryl Rowe
Production controller: Jess Lovell

Any URLs contained in this publication were checked for currency during the production process. Note, however, that the publisher cannot vouch for the ongoing currency of URLs.

First edition published in 2008.

Acknowledgements
Cover photograph: PhotoNewZealand / Julian Apse.

For product information and technology assistance,
in Australia call **1300 790 853**;
in New Zealand call **0800 449 725**

For permission to use material from this text or product, please email **aust.permissions@cengage.com**

National Library of New Zealand Cataloguing-in-Publication Data
Mathews, Roy.

Understanding science / Roy Mathews. 2nd ed.

Previous ed.: Melbourne, Vic. ; North Shore, N.Z. : Nelson Cengage Learning, c2008.
ISBN 978-0-17-018952-1
1. Science—Textbooks. 2. Science— Problems, exercises, etc.
I. Title.

507.6—dc 22

Cengage Learning Australia
Level 7, 80 Dorcas Street
South Melbourne, Victoria Australia 3205

Cengage Learning New Zealand
Unit 4B Rosedale Office Park
331 Rosedale Road, Albany, North Shore 0632, NZ

For learning solutions, visit **cengage.co.nz**

Printed in Australia by Ligare Pty Limited.
6 7 8 9 10 11 12 21 20 19 18 17

CONTENTS

GENETIC VARIATION

SPECIFIC LEARNING OUTCOMES

✓ Recognise DNA as the genetic material of most living things.

✓ Describe the process of DNA replication.

✓ Explain the roles of DNA in carrying instructions to the next generation, and in determining phenotype.

✓ Describe the relationship between chromosomes, alleles, genes and DNA.

✓ Discuss the importance of mitosis and meiosis in the growth and reproduction of living organisms.

✓ Describe genetic variations, and some of the factors that produce variation.

✓ Describe the significance of sexual reproduction in producing a new mix of alleles.

✓ Discuss the advantages and disadvantages of sexual and asexual reproduction.

✓ Describe mutations, and explain how mutations cause changes in phenotype.

✓ Explain how rates of survival by various members of a group may depend on their different phenotypes.

✓ Discuss the importance of variation within populations in a changing environment such as pest infestation, disease, drought, flood.

✓ Explain how genotype determines phenotype.

✓ Explain patterns of inheritance involving simple monohybrid inheritance showing complete dominance, sex determination, possible genotypes, and phenotype ratios.

✓ Draw and interpret punnett square diagrams and pedigree charts to describe the inheritance of a particular trait and the inheritance of sex.

✓ Discuss some advantages and disadvantages of biotechnology.

ISBN: 978-0-17-018952-1

Cells are the building blocks of a living organism. All cells, no matter whether in plants or animals, contain certain structures in common.

Each animal or plant cell has a control centre, the **nucleus**. It contains the genetic material. The genetic material is packed in structures called **chromosomes**. Chromosomes are visible only in a dividing cell. Each species of animal and plant has its own characteristic number of chromosomes. Human beings normally have 46 chromosomes (23 pairs) in every body cell. A pair of matched chromosomes is called a **homologous pair**. You have 23 homologous pairs of chromosomes in each body cell, one set from your mother and one set from your father. Homologous pairs of chromosomes are identical in size, shape and the position of genes on them. The only exception to this is the sex chromosomes. Males have an X and a Y chromosome which are not identical in their size or the locations of genes on them. Females have two X chromosomes. **Gametes** (sex cells) have only half of the original number of chromosomes. **Sperm** and **eggs** of human beings have only 23 chromosomes in them, one from each homologous pair. The chart below gives you the number of chromosomes in other species.

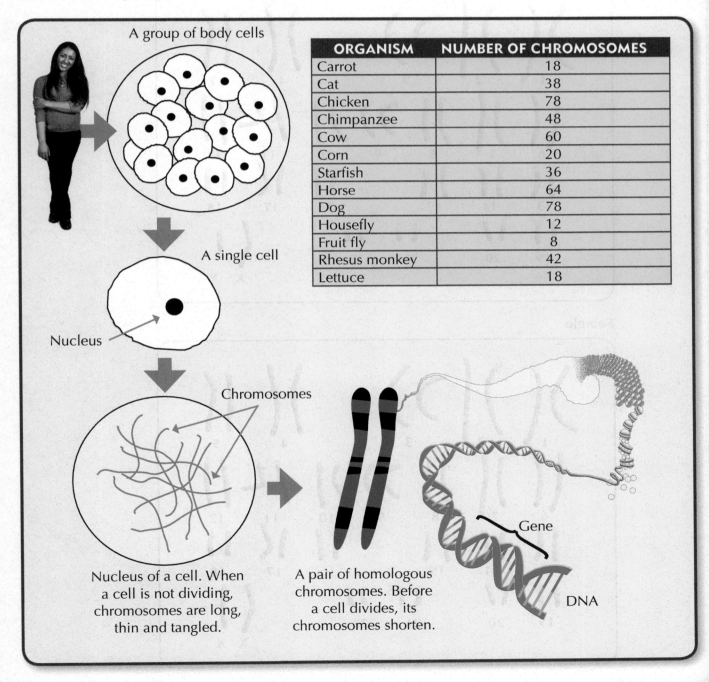

A group of body cells

A single cell

Nucleus

Chromosomes

Nucleus of a cell. When a cell is not dividing, chromosomes are long, thin and tangled.

A pair of homologous chromosomes. Before a cell divides, its chromosomes shorten.

Gene

DNA

ORGANISM	NUMBER OF CHROMOSOMES
Carrot	18
Cat	38
Chicken	78
Chimpanzee	48
Cow	60
Corn	20
Starfish	36
Horse	64
Dog	78
Housefly	12
Fruit fly	8
Rhesus monkey	42
Lettuce	18

ISBN: 978-0-17-018952-1

Chromosomes contain lengths of a chemical called **DNA,** wound around certain proteins. Every cell in your body carries exactly the same set of chromosomes, each made of the same specific chemical molecule, the DNA. The only exceptions to this are red blood cells (which do not have nuclei) and sex cells.

Genes, the basic units of heredity, are found on chromosomes. Each gene is a section of DNA that holds the instructions on how to build a specific protein. Humans have about 25,000 genes. Every cell has two versions of each gene, one inherited from each parent. Alternate forms of a particular gene are called **alleles**. We will learn about alleles later in this course.

The sum total of all the genes of an organism is called its **genome**. The complete set of chromosomes in the cell of an organism is called a **karyotype**. This is a diagram of chromosomes arranged in homologous pairs.

The karyotypes shown here are from human body cells, one male, one female. Each has 22 pairs of **autosomes** and a set of non-identical **sex chromosomes** (XY for males and XX for females). Autosomes are chromosomes that are not sex chromosomes.

Male

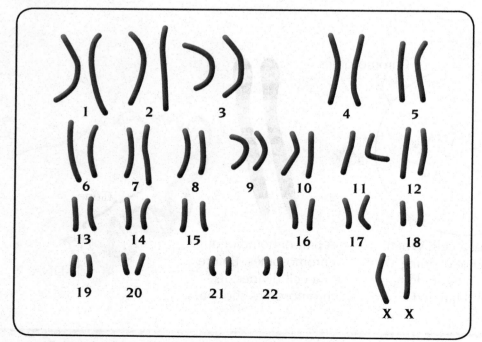

Female

ISBN: 978-0-17-018952-1

EXERCISES

A **The diagram below shows an animal cell and structures associated with it. Use it to answer the following questions.**

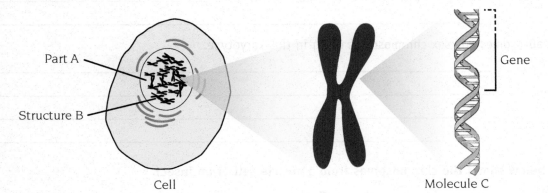

Part A

Structure B

Cell

Gene

Molecule C

1 Part A of the cell is called the 'control centre' of the cell. Name this part of the cell and explain why it is called the 'control centre'.

_____

_____

_____

_____

2 Structure B is found in the cell as homologous pairs. Name structure B.

_____

3 Explain what 'homologous pair' means.

_____

_____

_____

_____

4 Name molecule C.

_____

5 A small section of molecule C is labelled as 'gene'. Explain what a gene is.

_____

_____

_____

B **Given below is a human karyotype. Use this to answer the questions.**

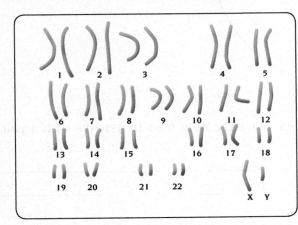

1 State how many pairs of chromosomes are pictured in this human karyotype.

_____

2 Chromosome make-up of an unfertilised human egg cell is different from this karyotype. Explain in what way the chromosome make-up of an unfertilised egg cell differs from this karyotype.

_____

_____

_____

ISBN: 978-0-17-018952-1

3 Explain from where a human body cell gets homologous pairs of chromosomes.

_____

_____

_____

_____

4 Describe how autosomes and sex chromosomes differ in this karyotype.

_____

_____

_____

_____

C **The diagram below shows the chromosomes from a muscle cell of an insect.**

1 State the number of chromosomes in:

   a a sperm cell produced by this insect. _____

   b a skin cell of this insect. _____

   c an egg cell produced by this insect. _____

   d a nerve cell of this insect. _____

2 In the space provided draw the chromosomes in a gamete produced by this insect.

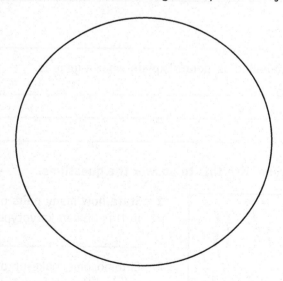

3 In the body cell of this insect chromosomes are in pairs. State the word used to describe such a pair of chromosomes.

_____

_____

ISBN: 978-0-17-018952-1

4 Write definitions for the following terms.

a Gamete

_____

_____

_____

b Karyotype

_____

_____

_____

c Genome

_____

_____

_____

d Allele

_____

_____

_____

e Autosomes

_____

_____

_____

C **Each body cell of a mouse contains 40 chromosomes.**

1 Write down the number of chromosomes found in each type of cell shown below.

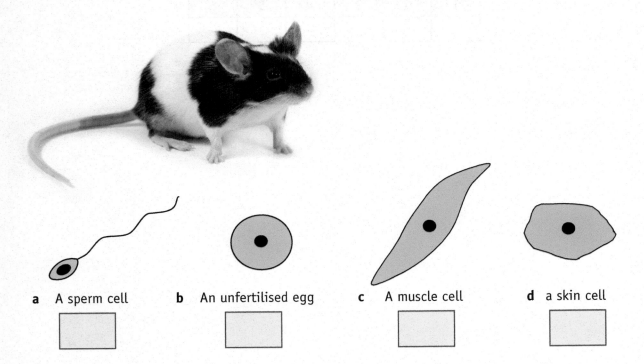

a A sperm cell

b An unfertilised egg

c A muscle cell

d a skin cell

D Answer the following.

1 Define the term 'chromosome'.

_____

_____

2 Define the term 'gene'.

_____

_____

3 Describe the link between chromosomes and genes.

_____

_____

_____

_____

E Unscramble the following words, then match them with their meanings or descriptions.

1 MOSSEMOCHOR _____

2 NEESG _____

3 SUNLECU _____

4 MOGNEE _____

5 GSHUOOMLOO _____

6 PARTYKOYE _____

A A picture that shows all the chromosomes found in the body cell of an organism.

B A pair of identical chromosomes.

C Structures that hold the genetic material of an organism.

D The hereditary units of an organism.

E Sum total of all genes found in an organism.

F The control centre of a cell.

Answers

1	2	3	4	5	6

ISBN: 978-0-17-018952-1

DNA (DEOXYRIBONUCLEIC ACID)

2

Deoxyribonucleic acid (**DNA**) holds the genetic information used in the development and functioning of every living organism (but some viruses have RNA instead of DNA as their genetic material). DNA holds the genetic information which directs the cell to construct specific proteins. Proteins make up the structure of the cell, and also the enzymes that control all biochemical reactions within the cell.

DNA is a **polymer** of simple repeating units called **nucleotides**. (A polymer is a large molecule made up of several repeating units called **monomers.**)

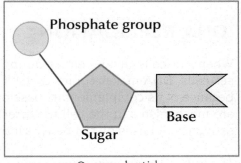

One nucleotide

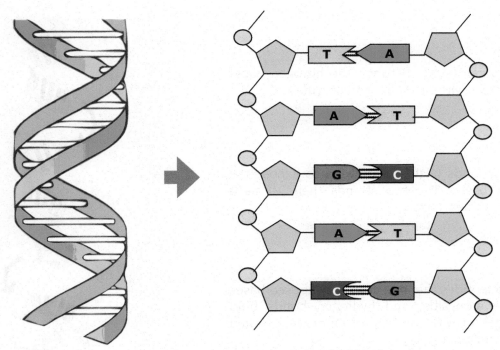

DNA shape: a double helix

Part of DNA that shows the pairing of bases

Nucleotides are the monomers of DNA. Each nucleotide is made up of three component parts, a nitrogen-containing **base**, a **deoxyribose sugar** molecule and a **phosphate group**. These nucleotides are arranged in such a way that it gives DNA its characteristic double helix shape, which is like a twisted ladder. The 'sides' of the ladder are made from alternating sugar and phosphate groups. The 'rungs' are made up of pairs of bases joined by weak **hydrogen bonds**.

There are four different types of base: adenine (**A**), guanine (**G**), cytosine (**C**) and thymine (**T**). Each type of base on one strand bonds with only one type of base on the other strand, A bonds with T and G bonds with C. This is called **complementary base pairing**.

The hydrogen bonds holding these bases together can easily be broken and re-formed. The order in which the bases are arranged in a gene determines the sequence of **amino acids** (amino acids are the building blocks of a protein) within a protein it codes for. The type of protein the DNA directs to synthesise in the cell makes unique 'traits' in organisms. The trait can be part of an organism's physical appearance such as a person's eye colour or it can be a person's resistance to diseases. Some of the traits a person has are produced by the interaction between the genes and the environment. For example, if a child who inherits genes for being tall is malnourished, he or she may not grow as tall as he or she could in a better environment that provides good nourishment.

ISBN: 978-0-17-018952-1

In a cell, the DNA is located inside the nucleus while the construction of protein occurs in the cytoplasm of the cell. The information coded on the DNA must be transferred into the cytoplasm for the amino acids to be assembled according to the genetic information coded in DNA. This is done by a molecule very similar to DNA called **ribonucleic acid** (RNA). Each three bases (a **codon**) code for one amino acid.

DNA REPLICATION

Whenever cells divide, either during growth, replacement of old cells or during the production of sex cells, DNA makes copies of itself. This is called DNA **replication**. DNA is copied very accurately because of its complementary base pairing. If it makes mistakes during the replication process there are mechanisms in the cell to correct this immediately. When the auto-correction system fails, the sequence of bases on the newly synthesised DNA changes. A DNA change is called **mutation**.

The following five steps summarise the process of DNA replication.

1   The hydrogen bonds between the nitrogen bases start to break with the help of an enzyme.

2   The two strands unzip and unwind to expose the unpaired bases. The two strands then act as a template for new bases.

3   'New' nucleotides that match to the exposed bases assemble alongside, and hydrogen bonds bind them together with the help of an enzyme (called DNA polymerase).

4   Phosphate and sugar join up to form the 'sides' of the new strand.

5   The two DNA molecules wind up and achieve double helix shapes. Each molecule is half 'old' material, half 'new'.

This method of DNA replication can be described as a **semi-conservative** method. The newly formed DNA molecules contain only half of the original DNA, so 50% of the original DNA is conserved. The process of DNA replication is not as simple as described above. The DNA does not start replicating at one end. It starts as small bubbles and replication forks are formed at each end of the bubble. The DNA then unzip in both directions until all the bubbles join up and two double helices are formed.

ISBN: 978-0-17-018952-1

A Answer the following.

1 Describe the shape of a DNA molecule.

_____

_____

2 Explain why a DNA molecule is described as a polymer.

_____

_____

3 Write down the base pairing rule.

_____

_____

4 The diagram below represents part of a DNA molecule. Complete the diagram by filling in the missing bases.

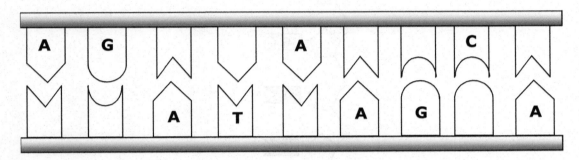

5 The diagram below represents a nucleotide. Label the three component parts of this nucleotide.

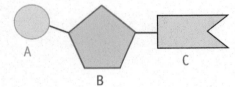

A _____

B _____

C _____

6 When a DNA molecule replicates, the following events happen (but not necessarily in this order).

A   Coiling brings about the formation of two helices.

B   Hydrogen bonds break, allowing the DNA strand to unzip.

C   Double helix unwinds.

D   New nucleotides take up positions on the templates.

Place the letters representing these events in the correct order.

7 Explain why DNA replication is described as semi-conservative.

_____

_____

_____

ISBN: 978-0-17-018952-1

8 DNA replication always happens before cell division. Explain why DNA replicates at this stage.

_____

_____

_____

_____

_____

_____

B **The diagram below shows a small section of one of the strands of a DNA molecule.**

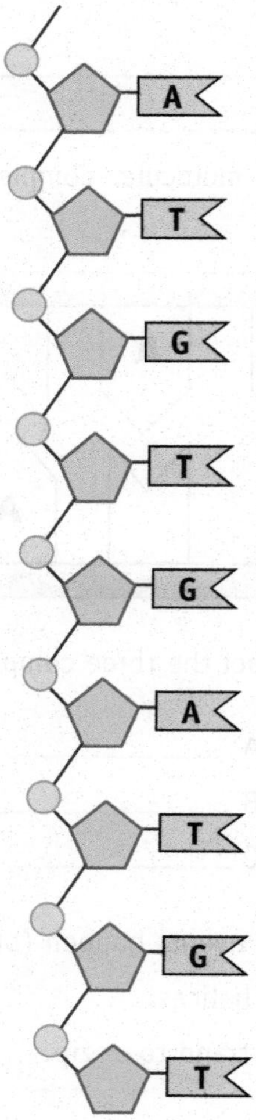

1 On the diagram above draw a circle to represent a nucleotide.

2 Place the following labels on the diagram:

a **nitrogen base**, a **sugar** and a **phosphate group**.

3 Explain how a gene acts as coded information for a protein.

_____

_____

_____

_____

_____

_____

_____

ISBN: 978-0-17-018952-1

4 State how many amino acids the above section of DNA could code for.

_____

5 In a long section of double-sided DNA with 700 bases total, it was found that adenine (A) occurred 200 times.

a State how many times thymine (T), cytosine (C) and guanine (G) occur.

_____

_____

_____

_____

b Explain how you worked out the above answer.

_____

_____

_____

_____

_____

6 Explain how the information coded on DNA gives an organism a particular characteristic such as eye colour in fruit flies.

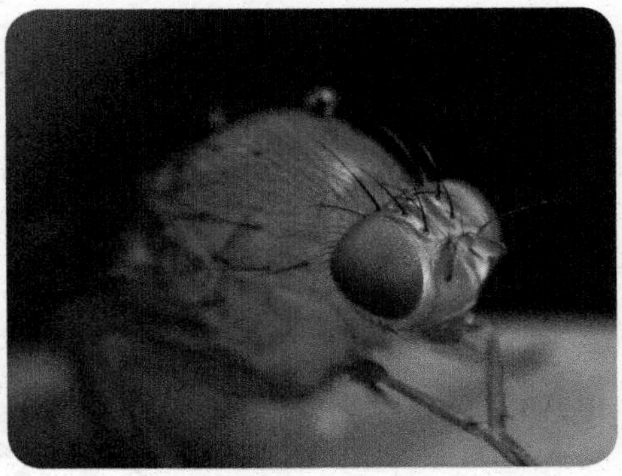

_____

_____

_____

_____

_____

_____

7 Explain how DNA is replicated and why information coded on the DNA has to be copied accurately. You must include the following points in your answer.

a A description of how DNA is replicated. You may use labelled diagrams to support your answer.

b An explanation of how this process maintains an accurate copying of genes.

c Relate reasons for the accuracy in DNA replication during cell division that occurs during growth of an individual organism.

_____

_____

_____

_____

_____

_____

_____

_____

_____

_____

_____

ISBN: 978-0-17-018952-1

3

CELL DIVISION

Cells of an organism multiply when it grows, produces gametes and when it replaces old or worn-out cells with new cells. There are two different types of cell divisions: **mitosis** and **meiosis**.

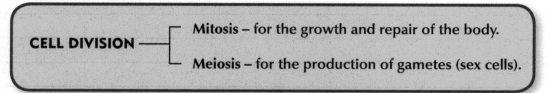

CELL DIVISION — Mitosis – for the growth and repair of the body.

Meiosis – for the production of gametes (sex cells).

The table below describes some major differences between mitosis and meiosis. Note: 'daughter' cells means 'new' cells (it has NOTHING to do with male and female).

MITOSIS	MEIOSIS
Occurs in all body cells except the cells in the reproductive organs that make gametes.	Occurs only in the reproductive organs (testes and ovaries in animals) before the formation of gametes (sperm and egg in animals).
Important for the growth and repair of the body and also for asexual reproduction.	Important for the sexual reproduction of an organism.
During mitosis one parent cell divides to form two identical daughter cells.	During meiosis one parent cell divides to form four daughter cells which may not be identical.
The number of chromosomes in the parent cell and the daughter cells remain the same. **Parent cell** **Daughter cells** Note: these chromosomes are shown in a very simple way.	Daughter cells get only half of the original chromosomes. **Parent cell** **Daughter cells** Note: these chromosomes are shown in a very simple way.
During mitosis, the homologous pairs of chromosomes do not pair up at any stage.	During meiosis, the homologous chromosomes pair up before separation.
No exchange of genes occurs between homologous pairs of chromosomes.	Genes are exchanged between homologous pairs of chromosomes before they separate.
Creates uniformity.	Creates variation.

ISBN: 978-0-17-018952-1

MITOSIS

Mitosis is a continuous process. For convenience, it can be divided into four stages with a preparatory stage at the start.

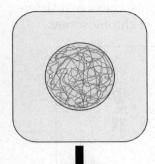

Between divisions:
This is a stage prior to cell division, although the cell is busy with other tasks. Chromosomes are thin and thread-like so they are not clearly visible. The cell prepares for cell division by replicating its DNA.

Stage 1: Double up
The nuclear envelope begins to disintegrate. Chromosomes thicken and shorten. Each has doubled and now becomes clearly visible as two strands or chromatids. **Spindle fibres** appear.

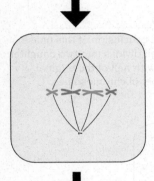

Stage 2: Line up
Chromosomes line up across the equator (middle) of the cell.

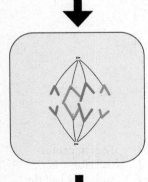

Stage 3: Split up
The chromatids of each chromosome separate and are dragged to the opposite poles of the spindle. The cell starts to split into two.

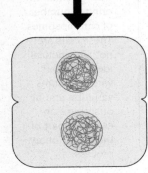

Stage 4:
The chromatids (now called chromosomes again) gain their long thread-like appearance and become invisible. A nuclear envelope appears around the two groups of chromosomes. The cytoplasm then divides to form two identical daughter cells.

ISBN: 978-0-17-018952-1

MEIOSIS

In animals, this type of cell division makes sex cells or gametes. It halves the number of chromosomes so that the resulting gametes receive only half the original number of chromosomes. When gametes fuse together and form a zygote during fertilisation, the normal number of chromosomes is regained. Thus meiosis helps to maintain the normal number of chromosome of a species. During meiosis, there are two successive divisions, called meiosis I and meiosis II. Note: **Haploid** refers to a cell with half the full number of chromosomes.

MEIOSIS I

Cell prepares for division.

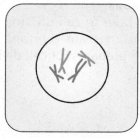

Chromosomes thicken and shorten.

Homologous pairs of chromosomes line up side-by-side at the equator of the cell.

Homologous chromosomes exchange pieces by a process called **crossing over**.

The two members of each pair move to the opposite poles of the spindle.

At the conclusion of this phase, the cell divides into two daughter cells with half of the original number of chromosomes.

MEIOSIS II

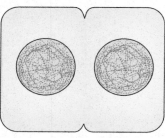

Two daughter cells formed from meiosis I undergo the second stages of meiosis.

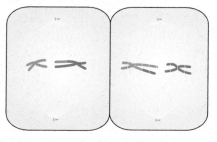

Chromosomes line up randomly at the equator and the chromatids begin to separate.

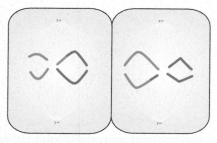

Chromatids separate and move towards the opposite poles with the aid of spindle apparatus.

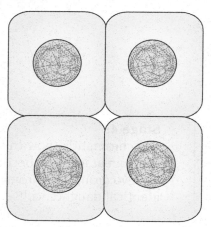

Four daughter cells form, each with only half the original number of chromosomes. These cells are not all genetically identical. This variety ensures variation among offspring – an essential part of the evolutionary process.

ISBN: 978-0-17-018952-1

EXERCISES

A The diagram below shows the stages of a type of cell division that happens at the growing bud of a stem cutting.

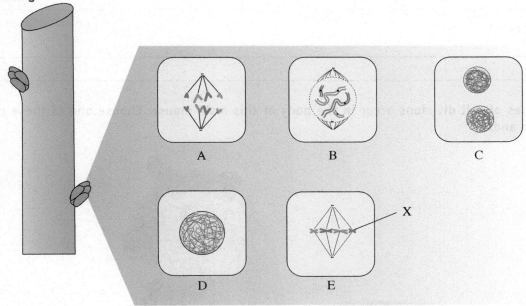

A   B   C

D   E   X

1 Place the letters of these diagrams in the right order in the boxes provided.

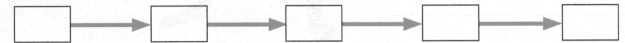

2 Name the type of cell division shown in the above diagrams.

_____

3 Give two reasons for your choice above.

_____

_____

4 Explain why this type of cell division happens especially in a growing bud.

_____

5 Name the structure labelled X on diagram E above.

_____

B The diagram below shows a process called crossing over happening during one of the early stages of meiosis. Study this to do the following.

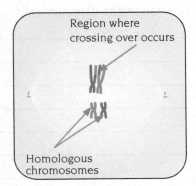

Region where crossing over occurs

Homologous chromosomes

1 Discuss the significance of this process.

_____

_____

_____

_____

2 Write down four important differences between mitosis and meiosis.

_____

_____

_____

_____

ISBN: 978-0-17-018952-1

3 Define the following terms:

  a Fertilisation:

_____

_____

  b Gametes:

_____

_____

C Two types of cell divisions occur in the body of this male mouse. Choose one of these types of cell division and:

Type of cell division:

```
┌─────────────────────────────┐
│                             │
│                             │
└─────────────────────────────┘
```

1 Describe exactly where this type of cell division takes place in the body of this mouse.

_____

_____

_____

2 Explain why this type of cell division reduces chromosome numbers to half.

_____

_____

_____

_____

_____

3 Discuss the importance of the number of chromosomes in the daughter cells produced by the type of cell division that you have chosen. You must state clearly if there is genetic variation in the daughter cells.

_____

_____

_____

_____

_____

_____

_____

_____

ISBN: 978-0-17-018952-1

D This is a diagram representing meiosis.

1 State an area in an adult human body you will find this type of cell division.

_____

_____

2 Explain why the number of chromosomes is halved.

_____

_____

_____

_____

_____

_____

_____

_____

_____

_____

_____

_____

_____

_____

_____

3 Cell division by meiosis produces genetic variation. Explain how meiosis increases genetic variety and its significance.

_____

_____

_____

_____

_____

_____

_____

_____

_____

_____

_____

_____

_____

_____

_____

_____

_____

_____

ISBN: 978-0-17-018952-1

E Complete the following chart by placing a tick (✓) or a cross (✗) in the correct box beside each statement.

DESCRIPTION	MITOSIS	MEIOSIS
Occurs at the growing tip of a root or a shoot of a plant.		
Occurs in the anther of a flower where pollen is made.		
Occurs in the testes of an animal where sperm cells are made.		
One parent cell divides to form two identical daughter cells.		
Chromosome number of the parent cell and the daughter cells remains the same.		
Daughter cells get only half of the original number of chromosomes.		
Exchange of genetic material occurs between the homologous pairs of chromosomes.		
This type of cell division helps when the body needs to replace old and worn-out cells.		

F The diagram below shows a flowering plant. Regions where active cell division occurs are circled.

1 Place MT in the circle where mitosis occurs (and meiosis never occurs) and place ME in the circle where meiosis occurs.

2 In the space provided, explain why this type of cell division is important at these regions of the plant.

This type of cell division is important here because

_____

_____

_____

_____

This type of cell division is important here because

_____

_____

_____

_____

ISBN: 978-0-17-018952-1

G The diagram below shows the life cycle of cattle. They have 60 chromosomes in their body cells. The circles represent the cells and the boxes represent the process involved in the production of cells.

1 Complete the diagram by writing down the numbers and words from the list below. Place chromosome numbers in each circle and the name of a process in each box.

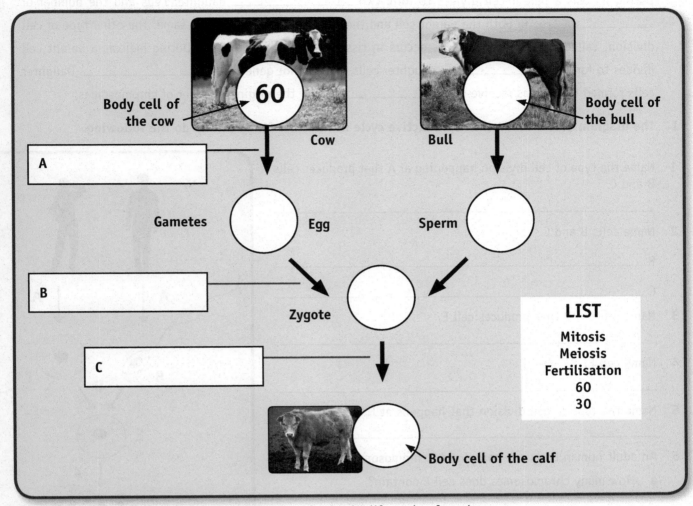

LIST

Mitosis
Meiosis
Fertilisation
60
30

2 Discuss the importance of processes **A**, **B** and **C** in the life cycle of cattle.

A _____

B _____

C _____

H The diagram below represents the chromosomes of a fly.

1 In the space provided below, draw in the chromosomes in a sperm cell and a muscle cell of this fly.

Sperm cell

Muscle cell

ISBN: 978-0-17-018952-1

2 Complete the following paragraph by filling in the blank spaces using appropriate words.

During the growth phase of an organism its cell _____. There are _____ types of cell divisions. The type of cell division responsible for growth is called _____. During mitosis a parent cell divides to form two _____ daughter cells and the number of _____ in both the parent cell and the daughter cells remain the same. The other type of cell division, called _____, occurs in tissues that makes gametes. During meiosis a parent cell divides to form _____ daughter cells that are not genetically _____. Daughter cells formed by meiosis receive only _____ of the original number of chromosomes.

I   The diagram below shows the reproductive cycle of humans. Study this to do the following.

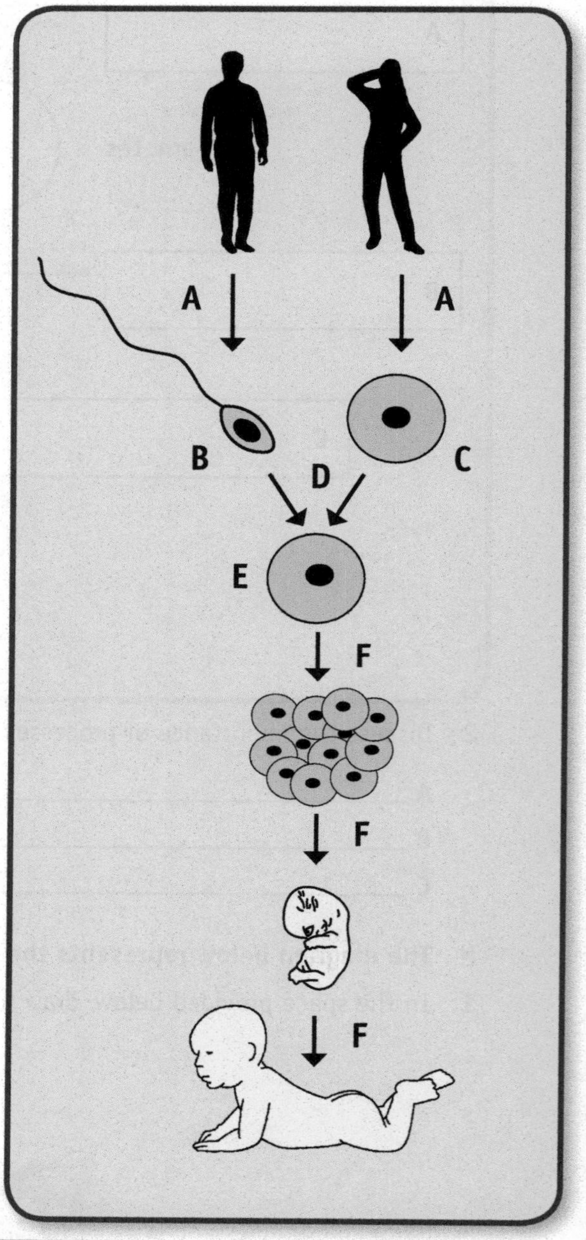

1 Name the type of cell division happening at A that produces cells B and C.

_____

2 Name cells B and C.

B _____

C _____

3 Name process D that produces cell E.

_____

4 Name cell E.

_____

5 Name the type of cell division that happens at F.

_____

6 An adult human body cell contains 46 chromosomes.

   a How many chromosomes does cell E contain?

   _____

   b How many chromosomes does cell C contain?

   _____

7 Chromosomes from a body cell of the mother and a body cell of the baby show some differences in their genetic make-up. Name two (or more) factors that cause this difference.

_____

_____

_____

_____

_____

_____

_____

_____

_____

ISBN: 978-0-17-018952-1

J **Domestic cats** (*Felis domesticus*) **have 38 chromosomes in their body cells.**

1 Complete the chart below by filling in the number of chromosomes present in each cell type.

CELL TYPE	NUMBER OF CHROMOSOMES
Muscle cell	
Sperm cell	
Skin cell	
Unfertilised egg	
Zygote	

2 Name two places in adult cats where meiosis happens.

_____

_____

3 Name two places in a kitten where mitosis happens.

_____

_____

K **Unscramble the following terms and then match them with their definitions or meanings.**

1 SITSOMI _____ **A** Organ that makes male gametes in animals.

2 SIMIOSE _____ **B** Organ that makes female gametes in animals.

3 TESMEAG _____ **C** A type of cell division that makes identical daughter cells.

4 ROYVA _____ **D** Type of cell division that makes gametes.

5 SETSET _____ **E** The process that fuses male and female gametes.

6 ARTNOITFLIIES_____ **F** Sex cells.

Answers

1	2	3	4	5	6

ISBN: 978-0-17-018952-1

Reproduction is vital for the survival and evolution of a species. Through reproduction an organism passes its genetic material or genes to the next generation. To ensure continuation of species, animals and plants use two methods of reproduction: sexual and asexual – also known as non-sexual reproduction.

SEXUAL REPRODUCTION

Sexual reproduction is a common mode of reproduction in most animals and flowering plants. It involves the fusing of two gametes, which in animals are produced as a result of meiosis. In most animals, it involves two parents, a male and a female. Each passes on half its genes to the offspring which thus possess a mix of genetic material that makes it different from both of its parents. This is called **genetic variation**.

As we have seen in the previous section, during meiosis, homologous chromosomes pair up at the equator of the spindle and one from each pair goes into each daughter cell randomly. As a result, the gametes produced by meiosis are not all identical. In fertilisation each gamete joins with another gamete, usually produced by another organism of the same species. The crossover that occurs during meiosis produces even greater variation. This variation is the major advantage of sexual reproduction. Offspring that are better adapted to new environmental conditions are more likely to survive and pass on their genes to the next generation.

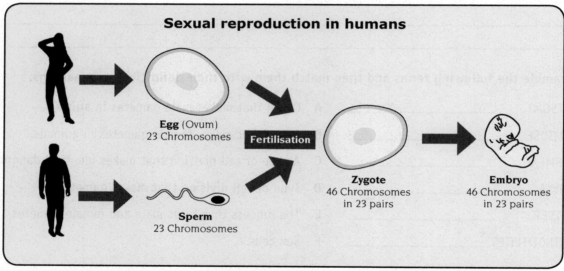

Sexual reproduction in humans

Egg (Ovum)
23 Chromosomes

Fertilisation

Sperm
23 Chromosomes

Zygote
46 Chromosomes
in 23 pairs

Embryo
46 Chromosomes
in 23 pairs

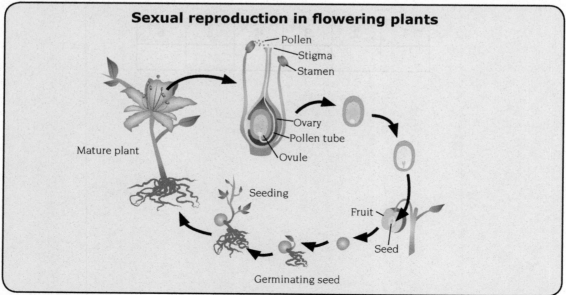

Sexual reproduction in flowering plants

Pollen
Stigma
Stamen

Ovary
Pollen tube
Ovule

Mature plant

Seeding

Fruit

Seed

Germinating seed

ISBN: 978-0-17-018952-1

ASEXUAL REPRODUCTION

Asexual reproduction is common in plants. It does not require two parents or the joining of two gametes. This means that one individual passes its entire set of genes to the next generation. As a result, offspring produced by asexual reproduction show no variation. Since all individuals are genetically identical none of them has any survival advantages in new environmental conditions. This makes them especially vulnerable to diseases or other natural disasters that can wipe them out.

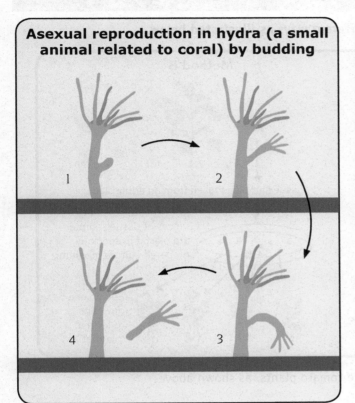

Asexual reproduction in hydra (a small animal related to coral) by budding

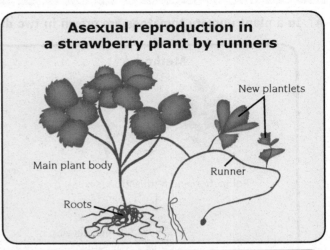

Asexual reproduction in a strawberry plant by runners

New plantlets

Main plant body

Runner

Roots

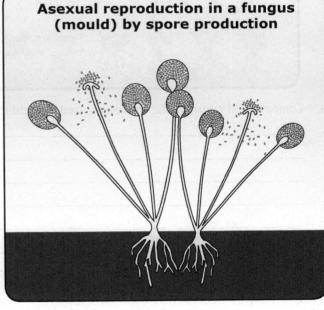

Asexual reproduction in a fungus (mould) by spore production

Eyes (buds) of a potato

Strawberry runner

African violet growing from piece of stem

In animals and plants, there are different forms of asexual reproduction.

Binary fission (e.g. bacteria) and budding (e.g. *Hydra*) are some of the types of asexual reproduction found in animals.

Vegetative reproduction and vegetative propagation of plants are also forms of asexual reproduction. Plants grown from cuttings (e.g. African Violets), runners (e.g. strawberry plants), tubers (e.g. potatoes), spores (e.g. fungi) are examples of asexual reproduction in plants.

ISBN: 978-0-17-018952-1

Asexual reproduction can be quick, with a few individuals producing hundreds of offspring in a short time. Another advantage: no mate is needed, so a new population can start from just one individual. Disadvantage: all these individuals are genetically identical; each is a **clone**.

EXERCISES

A **In a plant nursery tomatoes are grown in two different ways as illustrated below.**

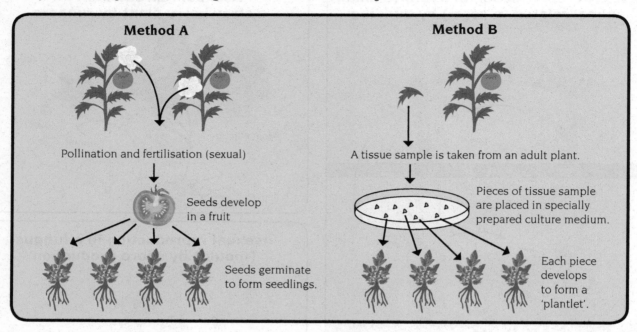

1 Compare and contrast the processes used to produce tomato plants, as shown above.

_____
_____
_____
_____
_____
_____

2 Discuss the advantages and disadvantages of producing tomato plants by the two methods illustrated above.

_____
_____
_____
_____
_____
_____
_____
_____
_____
_____

ISBN: 978-0-17-018952-1

B Septoria leaf spot is a common fungal disease that affects tomatoes. A gardener collected seeds from a tomato plant that has septoria leaf spot and raised seedlings for his next crop. He found some of the seedlings grown from the seeds are not affected by this disease.

1 Explain why some of the seedlings grown from seeds from an affected plant are resistant to septoria leaf spot disease.

_____

_____

_____

_____

_____

_____

2 A gardener wants to produce more tomato plants that have resistance to septoria. Suggest the best method of producing seedlings that have resistance to this fungal disease. Explain why you think this method will be 100% efficient in producing disease-resistant tomato plants.

_____

_____

_____

_____

_____

_____

_____

_____

C One of the methods commonly used to propagate plants involves the use of stem cuttings. A small piece of stem with leaves is taken and put into soil.

1 Explain one genetic advantages of growing plants from stem cuttings rather than from seeds.

_____

_____

_____

_____

_____

ISBN: 978-0-17-018952-1

2 One disadvantage of asexual reproduction is lack of genetic variation among offspring. Describe other disadvantages that could affect the population of plants such as strawberry that reproduce asexually.

_____

_____

_____

_____

_____

_____

_____

_____

_____

_____

3 A plant breeder has bred a new variety of camellia plant from seeds (sexual reproduction). To produce more of this camellia plant the breeder then used asexual reproduction.

Discuss the reasons why the breeder used sexual reproduction to produce the new variety of camellia plant and asexual reproduction to produce more plants.

In your answer you should:

- **link** mitosis AND meiosis to sexual and asexual reproduction

- **explain** how the genetic information of the parent are inherited in both sexual AND asexual reproduction

- **explain** why the breeder used both sexual AND asexual reproduction to produce a new variety of camellia plant.

_____

_____

_____

_____

_____

_____

_____

_____

_____

_____

ISBN: 978-0-17-018952-1

If you could look around and observe one particular characteristic of all students in your classroom, such as the shape of the ear lobes, you see variations. Some people have attached ear lobes while others have free ear lobes. Some variations are discrete or discontinuous while others are called continuous. Characteristics such the ability or inability to roll your tongue are examples of **discontinuous variation**. You can either be a tongue-roller or a non-roller. There is no in-between.

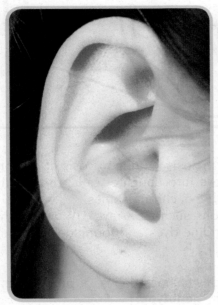

Attached ear lobe

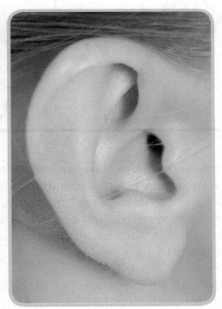

Free ear lobe

Characteristics such as your body height, weight and skin colour are examples of **continuous variation**. There is a wide range of heights, weights and skin colours. People are not tall or short, black or white.

Without variation, evolutionary changes will not occur in natural populations. The environment is always changing. Organisms with better adaptive features tend on average to survive longer than other members of the population with no favorable adaptations. This is called **natural selection**. Organisms with these favorable characteristics are more likely to reproduce and pass on these useful genes to the next generation.

The major sources of genetic variation are:
1 Mutations
2 Sexual reproduction

MUTATIONS

Mutations are permanent changes in the genetic material of an organism. This can range in size from a single DNA base to a large segment of a chromosome or even a change in the number of chromosomes.

Mutations occur randomly in a population. Only a small proportion of mutations are likely to be beneficial. Some are neutral, and most are harmful.

Mutations that occur in body cells are called **somatic mutations**. These are not passed on to offspring. Mutations that occur in reproductive cells can be passed on to offspring.

ISBN: 978-0-17-018952-1

Some of the causes of mutations are:

- Mistakes occurring when DNA replicates: during cell division. When DNA replicates, the bases may not copy correctly on to the newly formed DNA. A change in the DNA sequence may result in the cell synthesising a different protein.

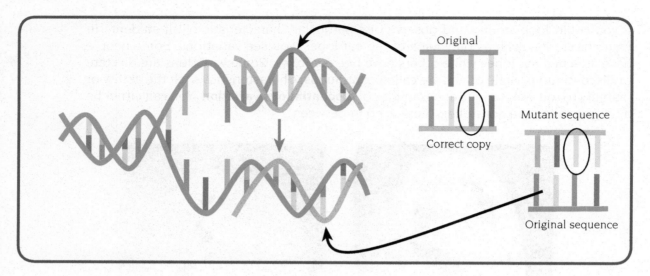

- Homologous chromosomes may not separate during meiosis. During meiosis, if one or more pairs of chromosomes fail to separate and move into daughter cells, the result would be cells with more or less than the normal number of chromosomes.
- External influences: Exposure to high energy radiations such as X-rays, UV rays, high temperature and chemicals such as nitrous acid, acridine dyes and formaldehyde can alter the bases on the DNA.

Some mutations produce a significant change in the bodily characteristics, or **phenotype.** The genes responsible for producing the phenotype are the **genotype**.

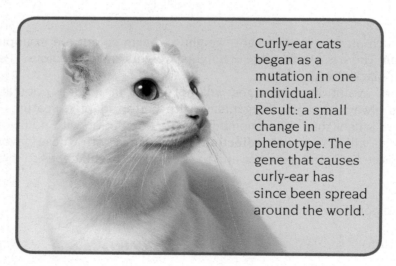

Curly-ear cats began as a mutation in one individual. Result: a small change in phenotype. The gene that causes curly-ear has since been spread around the world.

Some mutations can produce significant changes in phenotype. Mutant bacteria can develop resistance to antibiotics that normally kill them.

SEXUAL REPRODUCTION

Unlike asexual reproduction, sexual reproduction brings new combinations of genes into a population. As we have already seen, during meiosis, homologous chromosomes pair up together and then segregate randomly, moving into daughter cells. **Crossing over** which occurs during meiosis increases the genetic variation in gametes. The process of fertilisation then mixes these gametes from another individual that has also had its chromosomes segregated randomly.

ISBN: 978-0-17-018952-1

NATURAL SELECTION

Natural selection is the key mechanism of evolution. Some phenotypes (physical characteristics) produced by certain gene combinations have more chances of survival in a particular environment than others. Gradually, the frequency of these genes in a population increases and they replace the less favorable genes.

This caterpillar population consists of green coloured and brown coloured caterpillars.

The bird feeding on caterpillars finds it easier to pick up green caterpillars. Few of the green caterpillars survive to reproduce.

The number of genes responsible for producing this particular phenotype of green body colour is greatly reduced.

Caterpillars with more favourable adaptation, the brown body colour, survive longer and a higher proportion pass on their genes to their offspring. This process continues and the green colored caterpillars are eliminated from this caterpillar population.

EXERCISES

A Define the following.

1 Genetic variation

_____

_____

_____

_____

ISBN: 978-0-17-018952-1

2 Mutation

_____

_____

_____

_____

3 Phenotype

_____

_____

_____

_____

4 Genotype

_____

_____

_____

_____

B The diagram below shows a section of an original strand of DNA and a copy made by this DNA by undergoing replication.

1 Circle the area on the DNA below where a gene mutation has occurred during replication.

G	C	G	T	G	C	T	T	A	A	G	A	T	T	C	G

C	G	C	A	C	G	A	A	T	T	C	T	A	A	G	C

Original DNA

G	C	G	T	G	C	T	T	A	A	G	A	T	T	C	G

C	G	C	A	G	G	A	A	T	T	C	T	A	A	G	C

Copy made by replication

2 Explain how this type of mutation will tend to produce variation in the phenotype of an organism.

_____

_____

_____

_____

_____

_____

_____

ISBN: 978-0-17-018952-1

Given below is the karyotype of a person with the condition Down syndrome.

Down syndrome is caused by a chromosomal mutation.

3 Describe what the karyotype shows about the chromosomal abnormality of this person.

_____

_____

_____

_____

_____

_____

_____

_____

_____

4 Mutations that occur in body cells (somatic mutations) of sexually reproducing organisms will not be passed on to the next generation while mutations that occur in reproductive cells may be inherited. Explain this statement.

_____

_____

_____

_____

_____

_____

_____

_____

_____

_____

_____

_____

_____

ISBN: 978-0-17-018952-1

C Given below is an illustration that explains natural selection. The caterpillar population that feeds on leaves consists of green and brown caterpillars. The variation in their body colour is caused by a mutation. Birds feed on caterpillars.

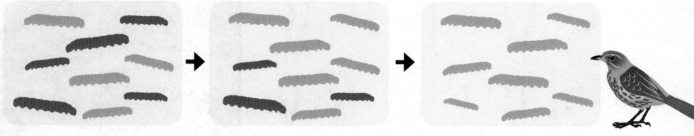

1 Explain natural selection using this illustration. You must consider the adaptive features of the caterpillars and their habitat in your explanation.

_____

_____

_____

_____

_____

_____

_____

_____

2 Cheetahs catch their antelope prey only if they can outrun the antelope in a short high-speed sprint. Explain how natural selection would have increased cheetah sprinting ability.

_____

_____

_____

_____

_____

_____

3 Explain how sexual reproduction produces variation among offspring in sexually reproducing organisms. You must include the following aspects in your discussion.

- Meiosis and gamete production.
- Fertilisation.

_____

_____

_____

_____

_____

_____

_____

_____

_____

ISBN: 978-0-17-018952-1

Genetics is the study of **heredity**. The foundation for this branch of science was laid by Gregor Johann Mendel (1822–84).

Hereditary **traits** or characteristics are determined by units called genes (Mendel called these 'factors'). We know that genes are sections of DNA that can direct the cell to make a particular protein.

A gene for a particular trait (for example height) can have different results or 'expressions' (for example tall or short). Alternative forms of a gene are called **alleles**. Letters of the alphabet can be used to represent the alleles (for examples T for tall and t for short).

TRAITS		ALLELES
Eye colour		Brown (B)
		Blue (b)
Ability to roll tongue		Tongue roller (R)
		Non-roller (r)
Shape of ear lobe		Free ear lobe (F)
		Attached lobes (f)
Middle finger hair		Hair present (M)
		No hair (m)

In the body cells of an organism, chromosomes are found in **homologous** pairs. Genes for a particular trait (feature) are found on both chromosomes exactly at the same location (**locus**). When an organism produces gametes, each gamete receives only one chromosome from the pair.

When alleles are identical, we say the gene pair is **homozygous** (for example TT for tall or tt for short). When an organism has two different forms of the allele, we call the gene pair **heterozygous** (for example Tt heterozygous tall). All **true breeding** or **pure breeding** organisms are homozygous for that particular trait.

In a heterozygous gene pair, the allele which expresses itself is called the **dominant** allele and the allele whose expression is hidden by the dominant allele is called the **recessive allele**. The dominant allele is usually represented by an upper-case letter and the recessive allele is represented by the corresponding lower-case letter. The genetic make-up of an organism (or simply the genetic formula) is called its **genotype** (for example TT, Tt, tt). The expression of an allele is called its **phenotype** (for example tall, short).

MONOHYBRID CROSSES

A cross between two organisms of the same species (male and female) to study the inheritance of a particular trait is called a **monohybrid cross.**

Below is an example of a simple monohybrid cross. This is one of the experiments conducted by Mendel on garden peas.

Two parent plants were selected. One was true breeding (homozygous) for round seeds and the other parent was true breeding for wrinkled seeds. The allele for round seed is R and the allele for wrinkled seed is r. Mendel crossed these plants by dusting the pollen from one plant onto the stigma of the other plant. He then collected all the seeds and planted them to study the first generation of offspring. He called them the **first filial generation** or **F1** generation. All the plants from this cross produced round seeds.

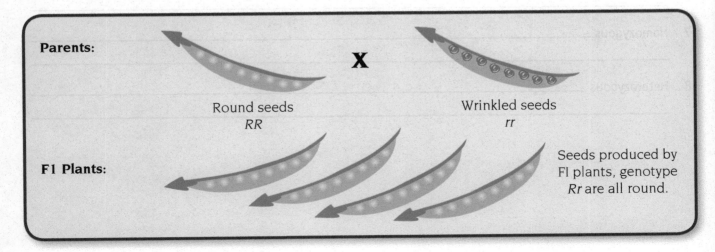

Parents: Round seeds RR   X   Wrinkled seeds rr

F1 Plants: Seeds produced by Fl plants, genotype Rr are all round.

ISBN: 978-0-17-018952-1

He then allowed the F1 plants (heterozygous round-seeded plants) to self-fertilise. The offspring produced by this cross, the **F2** generation, contained round-seeded and wrinkled-seeded plants in a definite ratio (3:1).

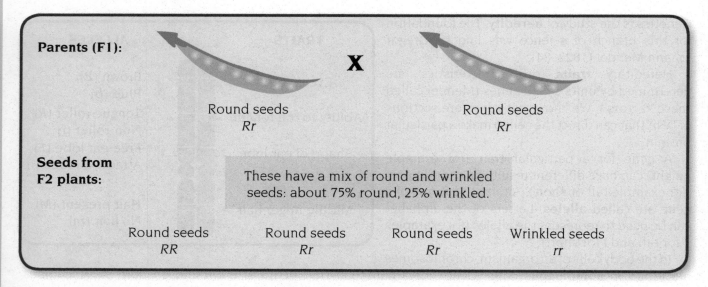

Parents (F1):

Round seeds
Rr

X

Round seeds
Rr

Seeds from F2 plants:

These have a mix of round and wrinkled seeds: about 75% round, 25% wrinkled.

Round seeds	Round seeds	Round seeds	Wrinkled seeds
RR	*Rr*	*Rr*	*rr*

EXERCISES

A   Give the meanings of the following words.

1   Gene _____

_____

2   Allele _____

_____

3   Genotype _____

_____

4   Phenotype _____

_____

5   Dominant allele _____

_____

6   Recessive allele _____

_____

7   Homozygous _____

_____

8   Heterozygous _____

_____

ISBN: 978-0-17-018952-1

B **Study the information below.**

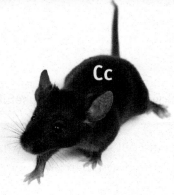

> The dark colour of this mouse is due to a dominant allele, *C*. The alternate allele is *c*. In its homozygous state *cc* produces a white coat colour.

1 State the genotype of this black mouse. _____

2 State the phenotype of this mouse. _____

3 State whether this mouse is homozygous or heterozygous. _____

4 This mouse has an allele for black colour and an allele for white colour but its coat colour is black. Give a reason for this.

_____

_____

5 State what the genotype of a homozygous black mouse would be.

_____

6 State what the genotype of a white mouse would be.

_____

C **The diagrams below show two fruit flies A and B. Fly A is heterozygous for its eye colour. Fly B is homozygous for its eye colour. Fly A has red eyes while fly B has brown eyes. The allele for red eye colour is R and the allele for white eye colour is r.**

Complete the chart below.

	FLY A	FLY B
Phenotype		
Genotype		
True breeding Yes/No		
Shows dominant/ recessive trait?		
Alleles in gametes	*R* or *r* (50%, 50%)	

FLY A

FLY B

ISBN: 978-0-17-018952-1

The following steps are very useful in solving genetic problems.

1. Find out the two parents involved in the cross.
2. Write down their phenotypes and genotypes. Remember there are two letters for each genotype.
3. Write down the gametes produced by each parent. Remember there is only one letter for each gamete.
4. Draw a **punnett square** and then place the gametes on the top and left-hand side of the punnett.
5. Combine the male and female gametes to fill in the boxes. This will give you the genotypes of the offspring. Always place the upper-case letters first.
6. Write down the phenotype underneath the genotype.
7. Write down the ratio as required.

Example

In pea plants, the allele for yellow seeds (**Y**) is dominant to the allele for green seeds (**y**). Find out the phenotype ratio of offspring produced by a cross between a heterozygous yellow-seeded plant with a green-seeded plant.

Step 1   Parents: heterozygous yellow-seeded plant and a green-seeded plant

Step 2
Phenotypes of parents: **Yellow seed** × **Green seed**

Genotypes of parents: Yy yy

Step 3 Gametes: Y y y y

Steps 4, 5 and 6

	Y	y
y	Yy Yellow	yy Green
y	Yy Yellow	yy Green

Step 7 Phenotype ratio: 2 Yellow : 2 Green.
This can be simplified to: 1 Yellow : 1 Green or 1 : 1

Note: since the green-seed parent can only produce gametes carrying y, it is not necessary to have a second y row in the punnett square. Whether you use two y rows or one y row, the final answer is the same.

ISBN: 978-0-17-018952-1

A Juan and Amber have been breeding rabbits. The diagram below shows the parents and offspring of one of their breeding experiments.

Parents
Both brown

Offspring
2 white
1 brown

In rabbits, coat colour is determined by alleles *B* and *b*. The allele for white coat *b* is recessive to the allele for brown coat *B*.

1 Complete the following genetic chart by writing the genotype of each rabbit. The genotype of the brown offspring has been completed for you.

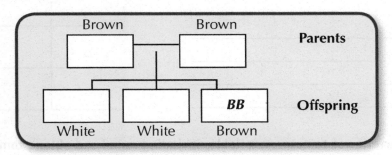

Brown   Brown   Parents

BB   Offspring

White   White   Brown

2 Juan and Amber want to breed more white rabbits from the same parents. Use the genotypes of the parents to complete the punnett square and predict the genotypes and phenotypes of the offspring.

3 What proportion of the offspring are expected to have white coat colour?_____

4 Explain why the predicted results of the cross above are different to the actual results of the mating shown in above in question 1.

_____

_____

_____

B White forelock, a patch of white hair at the front of the head, is a dominant genetic condition in humans. The allele for white forelock is *H* and normal hair is *h*. Maya is homozygous for white forelock. She married Caleb, who has normal hair. Hessa, their daughter, has the genotype *Hh*.

1 Complete the table below by filling in the genotypes and genotypes of Maya, Caleb and Hessa.

NAME	PHENOTYPE	GENOTYPE
Maya		
Caleb		
Hessa		*Hh*

2 *Olo, who has normal hair, married Hessa.* Complete the punnett square diagram to show the chance of different coloured hair in their children.

3 What is the chance that Hessa and Olo's first child will have white forelock?

_____

C Cystic fibrosis is a hereditary disease that affects many glands in the body. This is a recessive condition in humans. The allele for cystic fibrosis can be represented by *a* and the allele for unaffected people is **A**. Pita and Tania have one normal child but their second child was affected by this condition.

With the help of a punnett square diagram, explain how this occurred.

_____
_____
_____
_____
_____

D There are several varieties of tomatoes available, including round and oblong tomatoes. The allele for oblong shape **O** is dominant over the allele for round shape *o*.

In an experiment, a gardener pollinated the flower of a round tomato plant with the pollen from an oblong tomato plant. When the fruit ripened he collected 80 seeds. He planted these seeds and obtained the results as shown below.

Plants with oblong tomatoes 40%   Plants with round tomatoes 60%

1 State the genotypes of the parent plants.

a Round tomato _____

b Oblong tomato _____

2 Complete the punnett square to show the probability of different offspring in such a cross.

ISBN: 978-0-17-018952-1

3 State what proportion of the offspring are expected to have elongated tomatoes.

_____

4 What could the gardener do in order to make sure that no other pollens fertilise the flower during such crosses?

_____

_____

_____

E In sweet corn (Zea mays) the seed colour is determined by a single gene. Purple seed (*P*) is dominant over yellow seed (*p*).

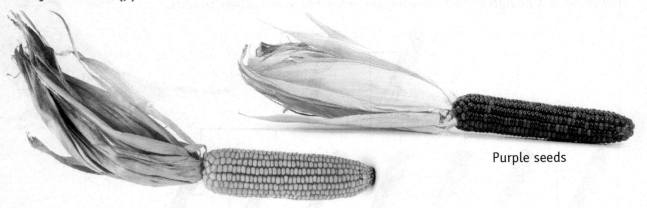

Purple seeds

Yellow seeds

A home gardener crossed a yellow seed corn plant with a purple seed corn plant. In the offspring he found both yellow seeds and purple seeds among the offspring.

1 State the genotype of the two corn plants he used for crossing.

  a Purple seed corn plant. _____

  b Yellow seed corn plant. _____

2 Complete the punnett square below to show the results of his cross.

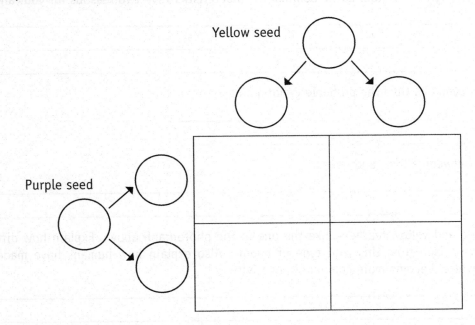

Yellow seed

Purple seed

3 State the likely phenotype ratio of the offspring in this cross.

_____

ISBN: 978-0-17-018952-1

4 A gardener bought some purple seeds from the garden centre. Explain what he could do to determine whether the corn seed he bought was homozygous or heterozygous. Use punnett squares to explain your answer.

_____

_____

_____

_____

_____

F Ming breeds budgies as a hobby. In her pet shop she has budgies of several colours - green, blue, grey and white. She keeps pedigree charts for each bird she sells. The following is a pedigree chart showing the inheritance of feather colours.

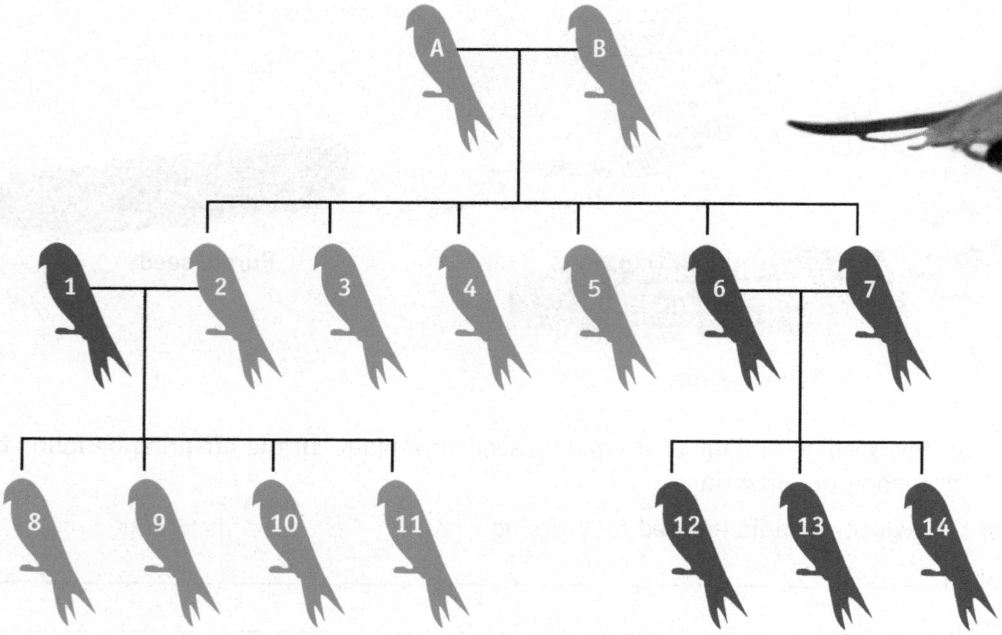

1 State which characteristic, green or blue, is the dominant characteristic. Give two reasons for your answer.

_____

_____

_____

2 Choose your own allele symbols. Give the probable genotype of

a bird A _____

b bird 6 _____

3 Explain the difference between alleles and genes.

_____

_____

4 Wild budgies have green and yellow feathers, like the one in the photograph above. Explain how different colours of budgies have arisen from this wild type of budgie. Also explain how humans have made new colours, like blue and white, become more common in captivity.

_____

_____

_____

_____

_____

ISBN: 978-0-17-018952-1

G In guinea pigs the allele for long hair (*h*) is recessive. The allele for short hair (*H*) is dominant.

1 Explain how a recessive allele can be 'expressed'.

_____

_____

_____

2 In a breeding trial, two short-haired guinea pigs were mated. The offspring were all short-haired. Explain why the genotypes of the parents could have been homozygous or heterozygous. Use punnett squares to support your answer.

_____

_____

_____

_____

_____

_____

_____

_____

_____

3 **Explain** how you could use a breeding trial to find out the genotype of a short-haired guinea pig. In your discussion you should LINK the following:

- a **description** of the possible genotype(s) of the short-haired guinea pig and the possible genotypes(s) of a breeding partner(s)
- a **description** of how the breeding trial would be carried out
- an **explanation** of how the outcomes in the offspring of the trial would determine the genotype of the short-haired parent guinea pig.

Use punnett squares to support your answer.

_____

_____

_____

_____

_____

_____

_____

_____

H Shown below is a pedigree chart for
the inheritance of hitchhiker's thumb.

Hitchhiker's thumb   Normal thumb

Black circles or squares indicate individuals with hitchhiker's thumb. White squares or circles show individuals with
normal thumbs. Circles represent females, squares males. Study this pedigree chart and answer the questions.

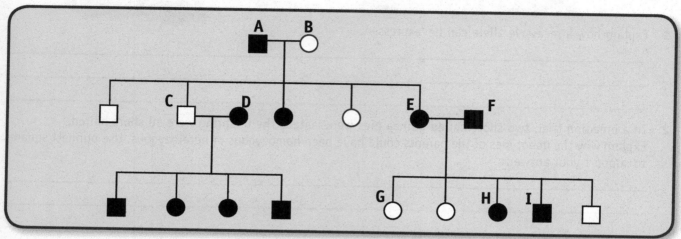

1 Which is the dominant characteristic shown in the above pedigree chart?

_____

2 Explain the reason why you think this trait is the dominant characteristic in this pedigree chart.

_____

_____

_____

_____

3 Using *H* to represent the dominant allele and *h* to represent the recessive allele, write down the genotype of
individuals A, B, E and F.

INDIVIDUAL	GENOTYPE
A	
B	
E	
F	

4 On the pedigree chart circle the two individuals who could be either homozygous or heterozygous for their trait.

5 Explain why the genotypes of these two individuals are in doubt.

_____

_____

_____

_____

In every body cell that has a nucleus, there are 23 pairs of chromosomes, carrying thousands of genes. Most pairs are 'matched', each one the same length as its homologous partner. One pair of chromosomes 'decides' whether an embryo will develop into male or female. Females have an XX pair, males have an XY pair of unequal length. We call these **sex chromosomes**. The other 22 pairs are known as **autosomes**.

The drawing below is based on a photo taken when the chromosomes had doubled, in preparation for cell division.

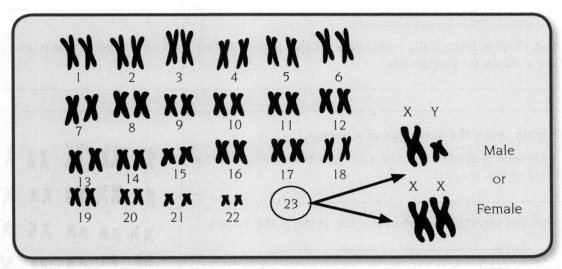

Meiosis occurs during gamete production. As a result of this each gamete gets only one sex chromosome. Since females have two X chromosomes, all eggs contain a single X chromosome, as well as 22 autosomes. Males produce 50% sperm cells with X chromosomes and 50% with Y chromosomes. The gender of each child depends on which sperm joins the egg during fertilisation.

Study the illustration below. Only the X and Y chromosomes are represented.

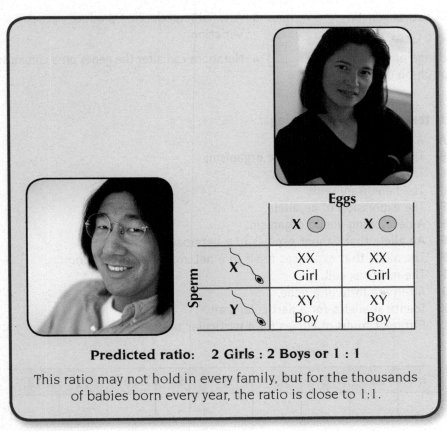

	Eggs	
	X ⊙	X ⊙
Sperm X	XX Girl	XX Girl
Y	XY Boy	XY Boy

Predicted ratio: 2 Girls : 2 Boys or 1 : 1

This ratio may not hold in every family, but for the thousands of babies born every year, the ratio is close to 1:1.

ISBN: 978-0-17-018952-1

EXERCISES

A Piripi and Mata are expecting their fourth child. They already have three boys and no girls.

1 State what chromosome in Piripi's sperm would produce a boy. _____

2 Complete the punnett square below to explain your answer to question 1 above.

3 State what the chance is of Piripi and Mata's next baby being a boy.

_____

_____

_____

4 Mata says Piripi is biologically responsible for the gender of their children. Do you agree or disagree with her? Give a reason for your answer.

_____

_____

B The diagram shows the karyotype of a person.

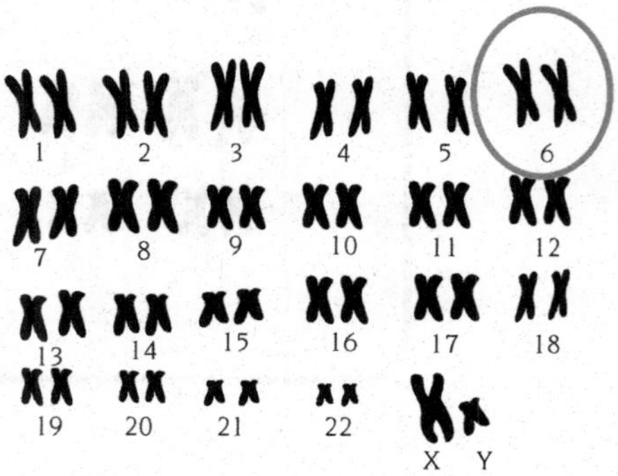

1 State what term is used to describe a pair of chromosomes like the one circled in this diagram.

_____

2 State what this karyotype tells you about the gender of this person.

_____

3 Give a reason for the above answer.

_____

_____

_____

C Write T (true) and F (false) beside each statement below.

☐ **1** Identical twins have identical genetic make-up.

☐ **2** During mitosis, exchange of genes occur between homologous chromosomes.

☐ **3** Asexual reproduction produces genetic variation.

☐ **4** Mutations can alter the genes on a chromosome.

D Match the following genetic terms with their meanings.

1	Genotype	**A**	A fertilised egg.
2	Phenotype	**B**	The genetic make-up of an organism.
3	Dominant	**C**	The female sex cell.
4	Recessive	**D**	The hereditary unit.
5	Allele	**E**	The expression of an allele.
6	Homozygous	**F**	A developing young organism.
7	Heterozygous	**G**	An allele that cannot express in a heterozygous condition.
8	Gene	**H**	The allele that expresses itself in a heterozygous condition.
9	Embryo	**I**	The male sex cell.
10	Sperm	**J**	Alternate form of a gene.
11	Egg	**K**	Identical alleles for a particular trait.
12	Zygote	**L**	Different forms of alleles for a particular trait.

Answers

1	2	3	4	5	6	7	8	9	10	11	12

ISBN: 978-0-17-018952-1

CLONING

A **clone** is a group of organisms produced from one parent (or ancestor), to which they are genetically identical. Identical twins are a natural clone. Cloning has become a controversial technology because of intentional cloning of animals such as sheep and cows.

A skin cell of an adult body, even though it contains the same genetic material as the zygote from which it originates, behaves differently. A skin cell, for instance, is a **differentiated** (specialised) cell. It can perform only certain jobs in your body because the genes responsible for other functions are inactivated during development. The egg cell or the **stem cells** (cells of embryos in their early stages) are **undifferentiated** (unspecialised) cells. They can develop to form a complete organism.

In one early experiment scientists cloned a frog by using a technique called nuclear transfer. In this method, the nucleus of an unfertilised egg of a frog was removed by UV radiation. The nucleus from a differentiated cell (such as skin cell) of another frog was then inserted into the egg cell. The egg cell now behaved like a fertilised egg with a full set of chromosomes. The frog produced by this experiment was identical to the frog that donated the nucleus.

Dolly, the first cloned sheep, was produced by a similar cloning technique. Scientists at the Roslin Institute in Scotland replaced the nucleus of an unfertilised egg cell with the nucleus from an udder cell of another sheep. The embryo was then inserted into the host mother (surrogate mother). After more than 200 tries, Dolly was born.

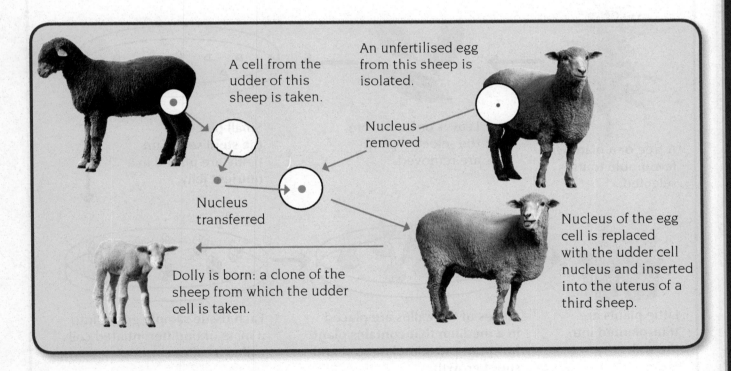

A cell from the udder of this sheep is taken.

An unfertilised egg from this sheep is isolated.

Nucleus removed

Nucleus transferred

Nucleus of the egg cell is replaced with the udder cell nucleus and inserted into the uterus of a third sheep.

Dolly is born: a clone of the sheep from which the udder cell is taken.

Cloning has advantages and disadvantages. Farming can be more productive by replicating the best animals, although this can also be achieved by traditional selective breeding methods. Scientists may be able to recreate extinct animals such as moa and dinosaurs by using the preserved genetic material in their fossils. But we cannot recreate an extinct environment.

However, cloning produces identical individuals, which reduces genetic diversity.

ISBN: 978-0-17-018952-1

TISSUE CULTURE

Unlike animal tissues, most plant **tissues** (clusters of cells) have the ability to develop into new plants. **Meristematic** tissues (growing parts of the plant such as buds) are ideal for such growth. Other parts like leaf and stem can also be used for tissue culture. The tissue culturing technique can be summarised in five steps:

1   Small pieces of leaves, stem or buds are cut off from a selected plant.
2   The tissue samples are sterilised by using disinfectants.
3   The tissue samples are then cut into small pieces and placed in a nutrient agar jelly. Each sample will then grow into a group of undifferentiated cells called a **callus**.
4   Small pieces of the callus are then placed in another solution (medium) that contains plant hormones. Different concentrations of these hormones stimulate growth.
5   Little plants are then transplanted into the soil.

Advantages of tissue culture

- Large numbers of plants can be produced from a small piece of tissue.
- All the plants produced by this method are genetically identical in all aspects. They all carry the traits of the parent plant.
- This can be quicker than raising plants from their seeds or cuttings.
- These plants are grown in the laboratory where viral or bacterial infections are controlled.

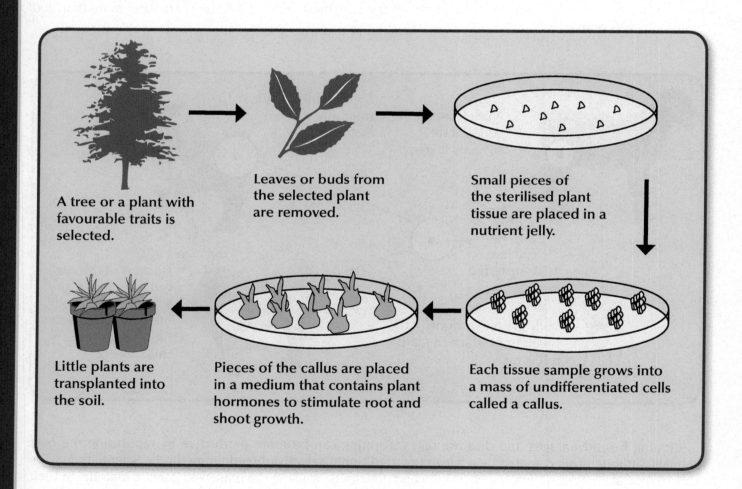

A tree or a plant with favourable traits is selected.

Leaves or buds from the selected plant are removed.

Small pieces of the sterilised plant tissue are placed in a nutrient jelly.

Little plants are transplanted into the soil.

Pieces of the callus are placed in a medium that contains plant hormones to stimulate root and shoot growth.

Each tissue sample grows into a mass of undifferentiated cells called a callus.

ISBN: 978-0-17-018952-1

SELECTIVE BREEDING

Plant and animal breeders want to produce plants and animals that possess 'desirable' characteristics such as high crop yield, resistance to disease and high growth rate. This can be achieved by **selective breeding**: select the organisms that show desirable traits for breeding purpose so that they will pass on the favourable genes to the next generation. This is usually done by crossing two members of the same species which possess the desirable traits. For example, crossing a plant that has an allele for high growth rate with another plant of the same species that possesses an allele for high yield should produce at least some offspring that will show both these desirable characteristics. This cross will produce plants that are fast-growing as well as high yielding.

Breeders continuously track which characteristics are possessed by each organism so when the breeding season comes once again, they can selectively breed the organisms to produce more favourable qualities in the offspring. This process of selecting the parents to produce offspring with desirable traits is called **artificial selection**. During the past 150 years selective breeding and improved animal husbandry have enabled cattle breeders to develop breeds which have a high efficiency for either milk or meat production.

GENETIC MODIFICATION

Genetic modification is a recent technology developed for altering the characteristics of plants and animals, in order to make them capable of making new substances or performing new or different functions. Like all organisms introduced into New Zealand, genetically modified organisms could have positive or negative effects on the environment, the economy and our society.

Genetic information of an organism is coded in its DNA. The sequence of bases on the DNA determines the protein it will help to assemble. A change in the base sequence will change the type of protein it helps to assemble, and thus the trait it produces will also change. This is the basic principle involved in **genetic engineering**. A gene that produces a favourable or good trait in one organism can be **spliced** into the DNA of another organism that cannot produce this feature. Advances in the field of Biotechnology also enable the scientists to identify faulty genes – genes that cause genetic disorders in humans and other organisms. In theory, these genes can be deleted and replaced with normal genes.

EXERCISES

A Answer the following.

1 Define 'clone'. _____

_____

2 'Propagation of plants by using cuttings is an example of cloning.' State whether you agree or disagree with this statement. Give a reason for your choice.

_____

_____

3 Most genetic modification experiments are conducted using stem cells obtained from embryonic tissues of animals. Explain how a stem cell differs from cells in an adult body.

_____

_____

_____

4 Summarise nuclear transfer technology as used in genetic engineering.

_____

_____

_____

ISBN: 978-0-17-018952-1

B **Use the information below to do the following.**

Cells are taken from a six-day-old embryo or from an adult cow.

A line of continuously reproducing cells is produced.

Each cell is put next to an egg which has had its nucleus removed.

Electric current fuses cells and the 'empty' egg. Cell nucleus containing DNA enters egg.

The embryos are inserted into host mothers that later give birth to the clones.

The egg now behaves like a fertilised egg, and each starts dividing.

1 If the first cell in the flow chart has 60 chromosomes, state how many chromosomes will be in each of the cells in the line of continuously dividing cells.

_____

2 Explain the genetic relationship between a cloned embryo and the host mother.

_____

_____

3 After an embryo has been formed, either by cloning or by sexual reproduction, it grows by cell division. State the name of this type of cell division. _____

_____

4 The normal chromosome number for cow is 60. State how many chromosomes would be in each sperm cell of a bull. _____

_____

5 Discuss one management or commercial advantage and one disadvantage to farmers producing cows that are identical to each other.

_____

_____

_____

C **Answer the following questions on tissue culture.**

1 Define 'meristematic tissue'.

_____

_____

_____

2 Give a reason why the tissue taken for culturing is washed in disinfectants.

_____

_____

3 Define a 'callus' in relation to tissue culture.

_____

_____

ISBN: 978-0-17-018952-1

4 Most plants reproduce sexually by producing flowers and seeds. Plants raised from seeds may show variation. What is the main cause of this variation?

_____

_____

_____

D There are different breeds of chicken. All of these chickens have the same common ancestor. They are descended from the Jungle Fowl, which can still be found in the wild. A wild Jungle Fowl might lay 20-30 eggs in a year. Today's hens each lay over 300 eggs a year on average.

1 Discuss how, with the help of humans, the modern breeds of chicken have changed from the ancestral type.

_____

_____

_____

_____

2 Some breeds of chicken are kept for their meat while others are farmed mainly for egg production. Explain how genetic engineering may help scientists to develop chickens that have both these favourable traits.

_____

_____

_____

_____

_____

_____

_____

E Complete the crossword puzzle below.

Across

4 A family tree showing the inheritance of a characteristic.
5 An alternative form of a gene.
6 Having two identical alleles for a given gene.
7 The genetic make-up of an organism.
9 The allele that determines the phenotype in a heterozygous condition.
10 The allele that has no noticeable effect on the phenotype in a heterozygous condition.

Down

1 A fertilised egg.
2 The expressed characteristics of an organism.
3 Having two different alleles for a given gene.
8 Sex cell.

ISBN: 978-0-17-018952-1

ACIDS AND BASES

SPECIFIC LEARNING OUTCOMES

✓ Describe the structure of an atom.

✓ Describe atomic number, mass number and electron arrangement of an atom.

✓ Describe what an isotope is.

✓ Draw diagrams of atoms to show the relative locations of three subatomic particles.

✓ Describe how an atom becomes an ion.

✓ Relate the charges of monatomic ions to their positions in the periodic table of elements.

✓ Use symbols or formulae of ions to write formulae of ionic compounds.

✓ Write word equations and balanced chemical equations for simple chemical reactions.

✓ Describe properties of acids and bases.

✓ Identify acids and bases using common indicators.

✓ Relate the acidity and alkalinity of a substance to its pH value.

✓ Describe neutralisation reaction.

✓ Describe reactions of acid with metals, metal oxides, metal hydroxides and metal carbonates.

✓ Identify the factors that affect the rate of a chemical reaction.

ISBN: 978-0-17-018952-1

PARTICLES

Pure substances made up of only one kind of atom are called **elements**. An atom is the smallest neutral particle of any chemical element. There are 111 known elements, which mean there are 111 different kinds of atoms. Atoms of two different elements are not identical, but they have certain things in common:

- All atoms have a central **nucleus**, which contains positively charged particles called **protons** and neutral particles called **neutrons**. Hydrogen is an exception. Most hydrogen atoms have no neutron in the nucleus, and only one proton.

- All around the nucleus there are negatively charged particles called **electrons**. These are arranged in different energy levels or **orbits** (also known as **shells**).

- Protons, neutrons and electrons are called **subatomic particles**.

- An atom is neutral because in an atom the number of positively charged protons is the same as the number of negatively charged electrons.

- The mass of an atom is mainly due to the particles inside the nucleus (protons and neutrons).Electrons have mass less than 1/1000 of proton mass.

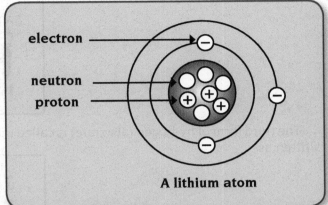

A lithium atom

ATOMIC NUMBER

The atomic number of an atom is the total number of protons in the nucleus.

Atoms of different elements have different atomic numbers.

For example the atomic number of a lithium atom is 3 (see the diagram above); The atomic number of a calcium atom is 20.

MASS NUMBER

The mass number of an atom is the total number of protons and neutrons present in the nucleus of an atom. For example in a lithium atom there are 3 protons and 4 neutrons, so the mass number of lithium is 7.

In a periodic table of elements the atomic number and mass number of an element are given.

atomic number = proton number

proton number = electron number

mass number = protons + neutrons

neutron number = mass number – atomic number

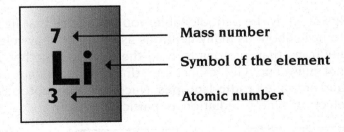

ISOTOPES

Isotopes are atoms of the same element with the same atomic number but a different mass number. Since they have the same atomic number, the number of protons is also the same. The reason why they have different mass numbers is because the number of neutrons in their nucleus is different.

ISBN: 978-0-17-018952-1

For example, the element hydrogen has three isotopes. The common form of hydrogen has no neutrons. The symbol for this common form of hydrogen is:

A rare form of hydrogen is called *deuterium*. It has one neutron and its symbol can be written as:

The third form of hydrogen (also rare) is called *tritium*. It has two neutrons and its symbol can be written as:

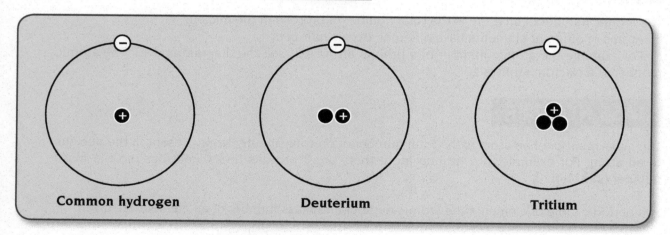

| Common hydrogen | Deuterium | Tritium |

If you study the periodic table, you will notice that light elements (smaller mass numbers) tend to have about as many neutrons as protons. Heavy elements (bigger mass numbers) have more neutrons than protons to hold the nucleus together. Atoms with a few too many neutrons, or not quite enough are unstable. Unstable atoms are radioactive. Some isotopes are radioactive and are called **radio-isotopes**. Nuclei of radio-isotopes are unstable. They decay by 'spitting out' electromagnetic radiation, or particles, or both.

 and  and  are three isotopes of carbon.

The first two are stable isotopes, while the third one is a radio-isotope.

ISBN: 978-0-17-018952-1

EXERCISES

A **The diagram below shows a chlorine atom. Study this and then do the following.**

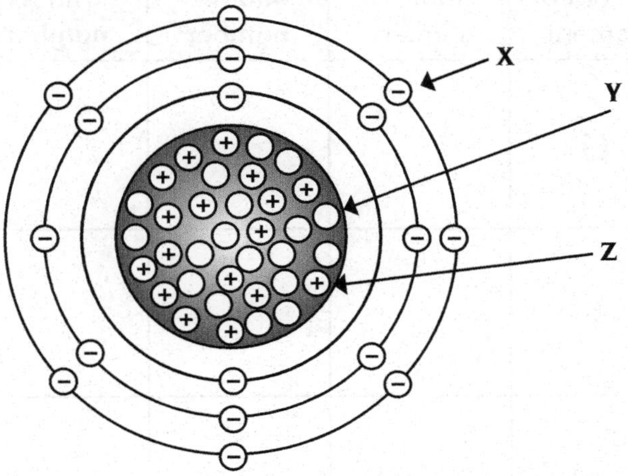

1 Name the particles labelled X, Y and Z.

X _____

Y _____

Z _____

2 State the atomic number of chlorine (Cl).

_____

3 State the mass number of Cl. _____

4 State how many protons Cl has. _____

5 State how many neutrons Cl has. _____

6 State how many electrons Cl has. _____

B **Complete the following sentences.**

1 The smallest neutral particle of an _____ is called an atom.

2 Pure substances made up of only one _____ of atom are called elements.

3 Mass of an atom is mainly due to _____ and neutrons present in its nucleus.

4 Three subatomic particles present in atoms are protons, _____ and _____.

5 Neutrons are neutral particles because they have almost no _____.

6 If the atomic number of an atom is 18 and its mass number is 40, then it must have _____ neutrons in its nucleus.

7 In a neutral atom, the number of proton is always equal to the number of _____.

C **Below is one of the boxes from the periodic table of elements. Study this, then answer the following questions.**

31
P
15

1 Name the element represented by this symbol. _____

2 Give the atomic number of this element. _____

3 Give the mass number of this element. _____

4 State how many protons an atom of this element has in one atom. _____

5 State how many electrons an atom of this element has in one atom. _____

6 State how many neutrons an atom of this element has in one atom. _____

D **Write T (true) or F (false) beside each statement.**

☐ **1** The atomic number of a carbon atom is the same as the atomic number of a calcium atom.

☐ **2** Electrons are the negatively charged particles found in an atom.

☐ **3** A calcium atom has 20 protons and its atomic number is 20.

☐ **4** A neutral aluminium atom has 13 electrons so it must have 13 protons also.

☐ **5** An atom has 17 electrons and its mass number is 35, so it has 18 neutrons.

☐ **6** Among the subatomic particles, electrons are the heaviest.

ISBN: 978-0-17-018952-1

E Complete the table below.

Symbol of element	Name of element	Atomic number	Mass number	Number of protons	Number of neutrons	Number of electrons
$^{7}_{3}\text{Li}$						
$^{19}_{9}\text{F}$						
$^{20}_{10}\text{Ne}$						
$^{23}_{11}\text{Na}$						
$^{27}_{13}\text{Al}$						
$^{35}_{17}\text{Cl}$						
$^{40}_{18}\text{Ar}$						
$^{39}_{19}\text{K}$						
$^{40}_{20}\text{Ca}$						
$^{64}_{29}\text{Cu}$						
$^{65}_{30}\text{Zn}$						

ISBN: 978-0-17-018952-1

F **Use the information below to answer the questions.**

> Chlorine has a wide range of isotopes, the two principal stable isotopes being ^{35}Cl (75.77% of a large number of Cl atoms) and ^{37}Cl (24.23%); they give chlorine atoms an average atomic mass of 35.4527 g/mol. Trace amounts of radioactive ^{36}Cl also exists in the environment.

1   Write down the atomic number and mass number of the most abundant isotope of chlorine. _____

2   Write down the atomic number and mass number of the radio-isotope of chlorine. _____

3   Describe what an isotope is. _____

_____

_____

4   ^{35}Cl and ^{37}Cl are neutral atoms. Explain what makes them neutral. Your explanation must include their atomic structure and electron number.

_____

_____

_____

5   Most periodic tables show the mass number (atomic mass) of chlorine as 35.45. Explain how chemists come up with this number.

_____

_____

_____

G **Given below are the symbols of three different isotopes of an element.**

1   State the name of the element.

_____

2   Explain the similarities and differences between atoms of these isotopes. Your explanation must include the number of sub-atomic particles and electron configuration.

_____

_____

_____

3   Complete this chart for the element Strontium:

Symbol of isotope	Atomic number	Mass number	Number of protons	Number of neutrons	Number of electrons
$^{84}_{38}Sr$					
$^{86}_{38}Sr$					
$^{87}_{38}Sr$					
$^{88}_{38}Sr$					

ISBN: 978-0-17-018952-1

In an atom, the negatively charged electrons are found outside the nucleus in different **shells** (also called **orbits** or **energy levels**). Smaller atoms have only one shell while bigger atoms have more than one electron shell. The way electrons are placed in their electron shells is called the **electron configuration** or simply electron arrangement.

If you follow the simple rules below, you will be able to write or draw the electron configuration for the first 20 elements in the periodic table.

Rules

(Note: these rules apply only to the first 20 elements.)

- The first shell, the one closest to the nucleus, can hold a maximum of 2 electrons.
- The second and third shells can each hold a maximum of 8 electrons.
- Any extra electrons are placed in the fourth shell.
- Inner shells are filled before any electrons are placed in the next shell.

The diagram shows the electron arrangement of a chlorine atom. A chlorine atom has 17 protons and 17 electrons. According to the above rule, the first shell can hold only 2 electrons. The second shell holds 8 electrons. Since this atom has only 17 electrons, the third and the last shell must hold the remaining 7 electrons.

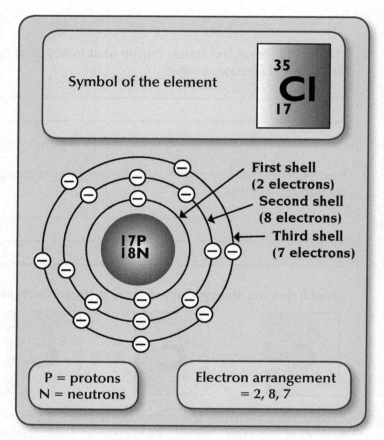

Symbol of the element — 35 Cl 17

First shell (2 electrons)
Second shell (8 electrons)
Third shell (7 electrons)

17P 18N

P = protons
N = neutrons

Electron arrangement = 2, 8, 7

EXERCISES

A Below are incomplete diagrams of four atoms. Study them and place electrons in their shells. Also write down the name and symbol of each element next to the diagram.

1

8P
8N

Name_____

Symbol_____

2

16P
16N

Name_____

Symbol_____

3

19P
20N

Name_____

Symbol_____

4

3P
4N

Name_____

Symbol_____

ISBN: 978-0-17-018952-1

B Answer the following.

The electron configuration of an atom is 2, 8, 8, 2. Its mass number is 40.

   a State the atomic number of this element. _____

   b State how many protons it has. _____

   c State how many electrons it has. _____

   d State how many neutrons it has. _____

   e Write down the name of this element. _____

   f Write down the chemical symbol of this element. _____

C Complete the table below. You may need to use a periodic table (see page 64) to complete this.

Name of element	Symbol	Atomic number	Mass number	No. of protons	No. of neutrons	No. of electrons	Electron arrangement
Hydrogen	H	1	1	1	0	1	1
	He		4	2			2
Lithium			7				2, 1
	Be	4			5		
					6	5	
			12				2, 4
Nitrogen				7	7		
		8			8		
	F		19				2, 7
Neon		10	20				
					12	11	
	Mg				12		2, 8, 2
Aluminium			27			13	
	Si				14	14	
			31	15			
				16	16		
			35				2, 8, 7
Argon		18			22		
	K		39	19			
			40	20			2, 8, 8, 2

ISBN: 978-0-17-018952-1

D Given below are symbols of six elements. Use the list to answer questions D1, 2, 3.

7 **Li** 3	35 **Cl** 17	40 **Ca** 20	27 **Al** 13	1 **H** 1	40 **Ar** 18

1 State which of the above elements fits each of these descriptions.

a an equal number of protons and neutrons _____

b a completely filled outer shell _____

c 2 electrons in its outer shell _____

d 4 shells _____

e only 1 shell _____

f the lowest mass _____

g the electron arrangement 2, 8, 3 _____

h no neutrons _____

i 18 neutrons _____

j only 1 electron _____

2 Which two elements have atoms that have

a the same mass numbers? _____ and _____

b 1 electron in their outer shell? _____ and _____

3 Atoms of which two elements need one more electron to achieve a complete set of electrons in their outer shell?

_____ and _____

E Below is the symbol of the element aluminium.

27 **Al** 13

1 Jan made a drawing as shown right to represent an aluminium atom, but she left out two kinds of particles found in the nucleus of this atom.

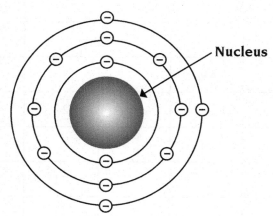

Nucleus

An aluminium atom

2 Write down the names of these two particles and their numbers in this table.

Name of particle	Number
a	
b	

ISBN: 978-0-17-018952-1

3 THE PERIODIC TABLE OF ELEMENTS

Elements are pure substances made up of *only one kind of atom.* The names and symbols of all known elements are listed in the periodic table.

In the periodic table, elements are placed in their order of atomic numbers. Hydrogen is the first element in the periodic table because its atomic number is 1.

There are 18 vertical columns in a periodic table. These vertical columns are called **groups**. All elements in one group show similar properties. Some of these groups have special names. For example Group 1 elements like Li, Na, K are called **alkali metals** and all of them react violently with water. Group 17 elements like Cl, Br, I are called **halogens**. Group 18 elements like He, Ne, Ar are called **noble gases** (also called inert gases) because they are reluctant to react with other elements. The horizontal rows of the periodic table are called **periods**.

In the periodic table (following page), the atomic number for each element is shown at the top left of the symbol, and the atomic mass or the mass number (larger number) is at the right. The atomic mass or the mass number is stated as a decimal number because it is an average of the various isotopes of an element. You may round this number for your calculations. For example, atomic mass of Li on this periodic table is 6.9; but it can be rounded to 7.

EXERCISES

A Below are the symbols of some chemical elements. Supply their names.

1 Na _____

2 K _____

3 He _____

4 Fe _____

5 Au _____

6 Hg _____

7 Ag _____

8 U _____

9 Ar _____

10 W _____

11 Sn _____

12 Sb _____

13 Pb _____

14 O _____

15 H _____

16 Ti _____

17 P _____

18 Pt _____

ISBN: 978-0-17-018952-1

The periodic table of elements

Key

	55.8
atomic number 26	
Fe	
iron	

relative atomic mass

gas at room temperature
liquid at room temperature
solid at room temperature
synthetic (does not occur naturally)

s block
p block
d block transition metals
f block lanthanides and actinides

1	2		3	4	5	6	7	8	9	10	11	12	13	14	15	16	17	18
1.0 **H** hydrogen 1																		4.0 **He** helium 2
6.9 **Li** lithium 3	9.0 **Be** beryllium 4												10.8 **B** boron 5	12.0 **C** carbon 6	14.0 **N** nitrogen 7	16.0 **O** oxygen 8	19.0 **F** fluorine 9	20.2 **Ne** neon 10
23.0 **Na** sodium 11	24.3 **Mg** magnesium 12												27.0 **Al** aluminium 13	28.1 **Si** silicon 14	31.0 **P** phosphorus 15	32.1 **S** sulfur 16	35.5 **Cl** chlorine 17	39.9 **Ar** argon 18
39.1 **K** potassium 19	40.1 **Ca** calcium 20		45.0 **Sc** scandium 21	47.9 **Ti** titanium 22	50.9 **V** vanadium 23	52.0 **Cr** chromium 24	54.9 **Mn** manganese 25	55.8 **Fe** iron 26	58.9 **Co** cobalt 27	58.7 **Ni** nickel 28	63.5 **Cu** copper 29	65.4 **Zn** zinc 30	69.7 **Ga** gallium 31	72.6 **Ge** germanium 32	74.9 **As** arsenic 33	79.0 **Se** selenium 34	79.9 **Br** bromine 35	83.8 **Kr** krypton 36
85.5 **Rb** rubidium 37	87.6 **Sr** strontium 38		88.9 **Y** yttrium 39	91.2 **Zr** zirconium 40	92.9 **Nb** niobium 41	95.9 **Mo** molybdenum 42	(98) **Tc** technetium 43	101.1 **Ru** ruthenium 44	102.9 **Rh** rhodium 45	106.4 **Pd** palladium 46	107.9 **Ag** silver 47	112.4 **Cd** cadmium 48	114.8 **In** indium 49	118.7 **Sn** tin 50	121.8 **Sb** antimony 51	127.6 **Te** tellurium 52	126.9 **I** iodine 53	131.3 **Xe** xenon 54
132.9 **Cs** caesium 55	137.3 **Ba** barium 56	57–71 lanthanoids	178.5 **Hf** hafnium 72	180.9 **Ta** tantalum 73	183.8 **W** tungsten 74	186.2 **Re** rhenium 75	190.2 **Os** osmium 76	192.2 **Ir** iridium 77	195.1 **Pt** platinum 78	197.0 **Au** gold 79	200.6 **Hg** mercury 80	204.4 **Tl** thallium 81	207.2 **Pb** lead 82	209.0 **Bi** bismuth 83	(209) **Po** polonium 84	(210) **At** astatine 85	(222) **Rn** radon 86
(223) **Fr** francium 87	(226) **Ra** radium 88	89–103 actinoids	(267) **Rf** rutherfordium 104	(268) **Db** dubnium 105	(271) **Sg** seaborgium 106	(272) **Bh** bohrium 107	(277) **Hs** hassium 108	(276) **Mt** meitnerium 109	(281) **Ds** darmstadtium 110	(280) **Rg** roentgenium 111								

57	58	59	60	61	62	63	64	65	66	67	68	69	70	71
138.9 **La** lanthanum	140.1 **Ce** cerium	140.9 **Pr** praseodymium	144.2 **Nd** neodymium	(145) **Pm** promethium	150.4 **Sm** samarium	152.0 **Eu** europium	157.3 **Gd** gadolinium	158.9 **Tb** terbium	162.5 **Dy** dysprosium	164.9 **Ho** holmium	167.3 **Er** erbium	168.9 **Tm** thulium	173.0 **Yb** ytterbium	175.0 **Lu** lutetium

89	90	91	92	93	94	95	96	97	98	99	100	101	102	103
(227) **Ac** actinium	232.0 **Th** thorium	231.0 **Pa** protactinium	238.03 **U** uranium	237 **Np** neptunium	(244) **Pu** plutonium	(243) **Am** americium	(247) **Cm** curium	(247) **Bk** berkelium	(251) **Cf** californium	(252) **Es** einsteinium	(257) **Fm** fermium	(258) **Md** mendelevium	(259) **No** nobelium	(262) **Lr** lawrencium

ISBN: 978-0-17-018952-1

B Below are the names of some elements. Write their chemical symbols.

1 Copper

2 Aluminium

3 Calcium

4 Magnesium

5 Silicon

6 Neon

7 Carbon

8 Chlorine

9 Oxygen

10 Lithium

11 Potassium

12 Sulfur

13 Zinc

14 Iodine

15 Manganese

16 Bromine

C Use the periodic table to do the following.

1 Write the symbols and names of the first three alkali metals.

_____

2 Write the symbols and names of the first three elements in group 2 of the periodic table.

_____

3 Write the symbols and names of the first four halogens.

_____

4 Write the symbols and names of the first four noble gases.

_____

5 Below are symbols and atomic numbers of four group 1 elements. Beside each symbol, write down its electron arrangement.

Symbol	Electron arrangement
H 1	
Li 3	
Na 11	
K 19	

6 Describe any similarity in the arrangement of electrons in atoms of these group 1 elements.

_____

_____

_____

_____

_____

_____

_____

ISBN: 978-0-17-018952-1

D Complete the crossword puzzle below.

Across
  2 The lightest element.
  5 This element is found in all organic compounds.
  6 This metal rusts in presence of water and air.
  8 This gaseous element is essential for most living things.
  9 A soft but dense metal.
 10 Compounds of this element are used in swimming pools.
 14 A liquid metal.
 15 This is a yellow-coloured solid non-metal.
 16 Compounds of this element can be found in your bones.

Down
  1 This non-metallic element exists as a liquid at normal room temperature.
  3 This metal burns with a brilliant flame.
  4 This metallic element has a reddish-brown colour.
  7 The most abundant gas in our atmosphere.
 11 The second lightest element.
 12 Alloy of this metal is used for making window frames.
 13 The name of this metal is same as its colour.

ISBN: 978-0-17-018952-1

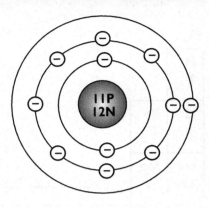

Ions are atoms or groups of atoms with electrical charges. There are positive ions and negative ions. How do neutral atoms get electrical charges? The diagram at the right shows a sodium atom. It has 11 protons and 11 electrons. Protons have positive charges and electrons have negative charges. These charges cancel each other out and we say the sodium atom is electrically **neutral**.

The outer shell of the sodium atom holds only one electron. This is not a stable arrangement. Every atom 'prefers' to have a full set of electrons in its outer shell. During chemical reactions, some atoms lose their outer electrons, while others gain one or more electrons to get a full set of electrons in their outer shell. There are some atoms which get a full outer shell by sharing electrons with other atoms.

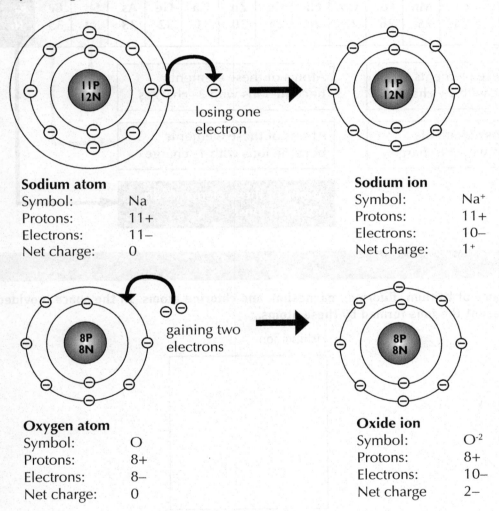

losing one electron

Sodium atom
Symbol:	Na
Protons:	11+
Electrons:	11−
Net charge:	0

Sodium ion
Symbol:	Na⁺
Protons:	11+
Electrons:	10−
Net charge:	1⁺

gaining two electrons

Oxygen atom
Symbol:	O
Protons:	8+
Electrons:	8−
Net charge:	0

Oxide ion
Symbol:	O^{-2}
Protons:	8+
Electrons:	10−
Net charge	2−

Atoms with 1, 2 or 3 electrons in their outer shell tend to lose these electrons and become positive ions. Atoms with 5, 6 or 7 electrons in their outer shell tend to gain electrons, and become negative ions. When non-metal atoms become negative ions, their names change. Chlorine atoms become chlor**ide** ions, bromine becomes brom**ide** and so on.

Atoms of most metallic elements lose electron(s) and become positive ions, while most non-metal atoms gain electron(s) to form negative ions. Atoms of group 1 alkali metals like Na lose one electron from the outer shell and form ions with 1+ charge. Atoms of most group 17 elements gain one electron and become ions with 1− charge. Elements between group 2 and 13 are called **transition metals**. Most of these elements form positive ions with 2+ charges. Atoms of group 18 elements do not form ions because they have stable electron arrangements with completely filled outer shells.

ISBN: 978-0-17-018952-1

An ion is an atom or group of atoms that has gained or lost electrons.
When an atom loses electrons, it becomes a positive ion. When an atom gains electrons, it becomes a negative ion.

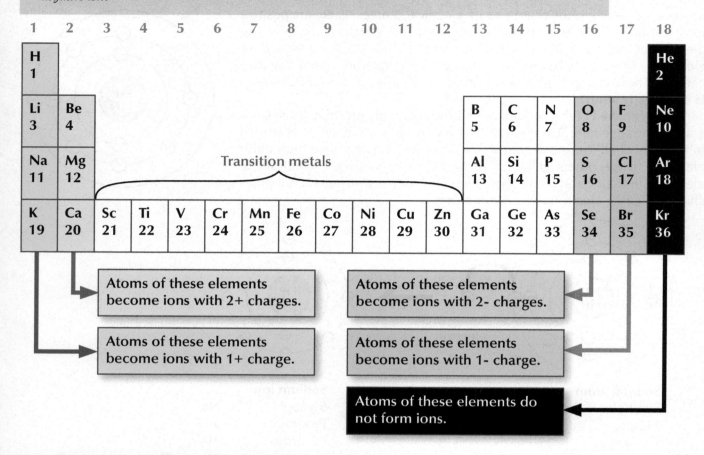

EXERCISES

A **Given below are diagrams of lithium, fluorine, magnesium and chlorine atoms. In the space provided draw diagrams to represent the ions formed by these atoms.**

1 Lithium atom

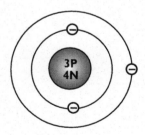

Lithium ion

2 Fluorine atom

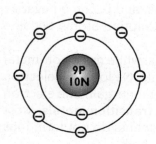

Fluoride ion

ISBN: 978-0-17-018952-1

3 Magnesium atom

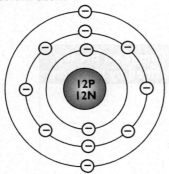

Magnesium ion

4 Chlorine atom

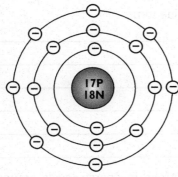

Chloride ion

B Complete the chart below.

Name	Symbol	Atomic number	Mass number	Number of protons	Number of neutrons	Number of electrons	Electron arrangement
Oxygen atom	O	8	16				
Oxide ion	O²⁻		16			10	2, 8
Chlorine atom					18		2, 8, 7
Chloride ion		17	35				
Potassium atom	K		39	19			
Potassium ion	K⁺				20	18	
Calcium atom			40				2, 8, 8, 2
Calcium ion		20	40				
Sulfur atom	S			16	16		
Sulfide ion			32		16		2, 8, 8

C Use your knowledge of ions and electron arrangements to explain the following.

1 Atoms of most group 17 elements form negative ions with a single negative charge.

_____

_____

2 Most group 18 elements are very unreactive inert gases.

_____

_____

ISBN: 978-0-17-018952-1

D Magnesium and sulfur form ions with different charges.

1 Complete the chart below.

Element	Group number	Electron arrangement	Charge on the ion formed by the atom
Magnesium			
Sulfur			

E The electron arrangement of a chloride ion is 2, 8, 8. This ion has a negative charge.

1 Explain how a neutral chlorine atom becomes a negatively-charged chloride ion.

_____

_____

_____

_____

_____

_____

2 The diagram below represents a neutral chlorine atom ($^{35}_{17}Cl$). Place electrons in energy levels around the nucleus. You may use this symbol $\ominus$ to represent electron.

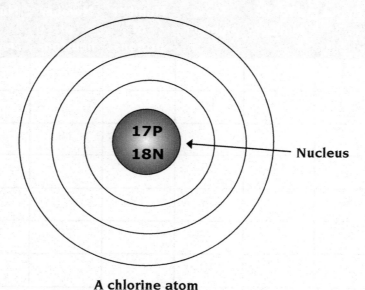

A chlorine atom

ISBN: 978-0-17-018952-1

5 GROUPING ELEMENTS

Elements in the periodic table can be grouped into **metals** and **non-metals**. Some elements like silicon and germanium show properties of both metals and non-metals, so they are called semi-metals or **metalloids**.

At ordinary room temperature, most elements remain in a solid state. Some elements are gases and some are liquids.

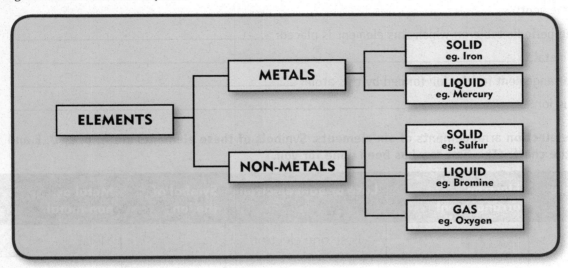

All metal elements are placed on the left-hand side of the periodic table and all non-metals are placed on the right-hand side of the periodic table. (H is the exception.)

1	2	3	4	5	6	7	8	9	10	11	12	13	14	15	16	17	18
H 1																	He 2
Li 3	Be 4											B 5	C 6	N 7	O 8	F 9	Ne 10
Na 11	Mg 12			■ Liquid		▨ Gas		▢ Solid				13 Al	14 Si	15 P	16 S	17 Cl	18 Ar
K 19	Ca 20	Sc 21	Ti 22	V 23	Cr 24	Mn 25	Fe 26	Co 27	Ni 28	Cu 29	Zn 30	Ga 31	Ge 32	As 33	Se 34	Br 35	Kr 36
Rb 37	Sr 38	Y 39	Zr 40	Nb 41	Mo 42	Tc 43	Ru 44	Rh 45	Pd 46	Ag 47	Cd 48	In 49	Sn 50	Sb 51	Te 52	I 53	Xe 54
Cs 55	Ba 56	to 57	Hf 72	Ta 73	W 74	Re 75	Os 76	Ir 77	Pt 78	Au 79	Hg 80	Tl 81	Pb 82	Bi 83	Po 84	At 85	Rn 86
Fr 87	Ra 88	to 89	Rf 104	Db 105	Sg 106	Bh 107	Hs 108	Mt 109	Ds 110	Rg 111							

Elements to the left of this line are metals (Except for H)

Elements to the right are non-metals

ISBN: 978-0-17-018952-1

EXERCISES

A The electron arrangement of an atom is 2, 8, 8, 1 and its mass number is 39. Complete the following facts about this element.

1 Atomic number: _____

2 Number of protons: _____

3 Number of neutrons: _____

4 Name of this element: _____

5 The side of the periodic table on which this element is placed: _____

6 Metal or non-metal? _____

7 The electron arrangement for the ion formed by this atom: _____

8 Symbol for this ion: _____

B Below are the electron arrangements of six elements. Symbols of these elements are A, B, C, D, E and F. Complete the chart. (The first row has been done for you.)

Symbol of the element	Electron arrangement	Lose or gain electron	Symbol of the ion	Metal or non-metal
Li	2, 1	Lose one electron	Li$^+$	Metal
F	2, 7			
Na	2, 8, 1			
O	2, 6			
Al	2, 8, 3			
Be	2, 2			

C Use the periodic table below to complete the following questions.

1 On the periodic table below, draw a line to separate metals from non-metals.

2 Use a highlighter to colour in all solid elements yellow.

3 Use a highlighter to colour in all gases green.

4 Use a highlighter to colour in all liquids (at room temperature) blue.

1	2	3	4	5	6	7	8	9	10	11	12	13	14	15	16	17	18
H 1																	He 2
Li 3	Be 4											B 5	C 6	N 7	O 8	F 9	Ne 10
Na 11	Mg 12											Al 13	Si 14	P 15	S 16	Cl 17	Ar 18
K 19	Ca 20	Sc 21	Ti 22	V 23	Cr 24	Mn 25	Fe 26	Co 27	Ni 28	Cu 29	Zn 30	Ga 31	Ge 32	As 33	Se 34	Br 35	Kr 36
Rb 37	Sr 38	Y 39	Zr 40	Nb 41	Mo 42	Tc 43	Ru 44	Rh 45	Pd 46	Ag 47	Cd 48	In 49	Sn 50	Sb 51	Te 52	I 53	Xe 54
Cs 55	Ba 56	to 57	Hf 72	Ta 73	W 74	Re 75	Os 76	Ir 77	Pt 78	Au 79	Hg 80	Tl 81	Pb 82	Bi 83	Po 84	At 85	Rn 86

ISBN: 978-0-17-018952-1

5 Write down the symbols and names of five common metals.

_____

_____

_____

_____

6 Write down the symbols and names of five non-metals.

_____

_____

_____

_____

_____

ISBN: 978-0-17-018952-1

COMPOUNDS

Compounds are pure substances made up of two or more kinds of atoms chemically bonded together. (Example: water, sugar, salt.) The atoms are chemically bonded together in exact proportions like 1:1 or 2:1.

A **formula** tells us the proportionate numbers of each kind of atom. In many cases, the atoms in a compound form a group. A group of two or more atoms is called a **molecule**.

Carbon dioxide is a compound. A molecule of carbon dioxide contains a carbon atom and two oxygen atoms and its formula is **CO$_2$**.

A CO$_2$ molecule

Study the following examples.

Sulfuric acid is a compound. Its formula is H$_2$SO$_4$. A formula like this tells us that one molecule of H$_2$SO$_4$ contains 2 hydrogen atoms, 1 sulfur atom, and 4 oxygen atoms.

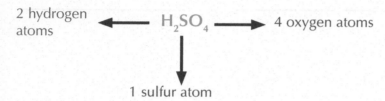

2 hydrogen atoms ← H$_2$SO$_4$ → 4 oxygen atoms

1 sulfur atom

The formula for calcium hydroxide is Ca(OH)$_2$. It means that a single calcium hydroxide molecule has 1 calcium atom, 2 oxygen atoms and 2 hydrogen atoms.

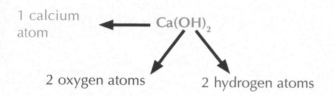

1 calcium atom ← Ca(OH)$_2$

2 oxygen atoms   2 hydrogen atoms

> When there is a number beside the bracket, you must multiply everything within the bracket by that number.

Sometimes you may find a number in front of the formula.

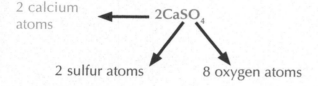

2 calcium atoms ← 2CaSO$_4$

2 sulfur atoms   8 oxygen atoms

> This formula tells you that there are 2 molecules of calcium sulfate present.

TYPES OF COMPOUNDS

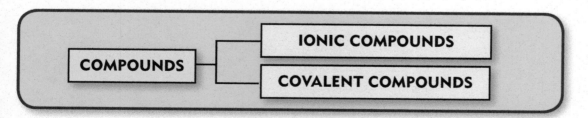

COMPOUNDS — IONIC COMPOUNDS

COVALENT COMPOUNDS

Elements react together to form compounds. When this happens, atoms bond together to form molecules with unique properties. Examples: the element chlorine is a poisonous gas (greenish), and sodium is a soft reactive metal (silvery). When atoms of these elements join together, a substance called sodium chloride (table salt) forms. Here, the sodium atom bonds with the chlorine atom to form sodium chloride. The chemical and physical properties of sodium chloride are very different from its 'parent'

ISBN: 978-0-17-018952-1

elements. Chlorine and sodium are poisonious, but sodium chloride is safe to eat.

How are chemical bonds formed between atoms by the interaction of electrons? As you have seen, atoms are most stable when their outer orbits have a full set of electrons. Very few elements have completely filled outer electrons (orbits). Atoms of most elements bond together either by receiving, donating or sharing electrons between them to get a full outer orbit. There are two major types of chemical bonding: ionic bonding and covalent bonding.

IONIC BONDING

In ionic bonding electrons are completely transferred from one atom to another. In this process, one atom will lose electrons while the other atom will gain electrons. These atoms become ions: positive or negative ions. Oppositely-charged ions are attracted to each other by strong electrostatic forces. The bonds between these oppositely-charged ions are called **ionic bonds**.

Sodium chloride (NaCl) is an **ionic compound**. A small crystal of sodium chloride contains billions of sodium ions and chloride ions (Na^+ and Cl^- ions). Every ionic compound contains positive ions and negative ions held together by the attraction of their opposite charges.

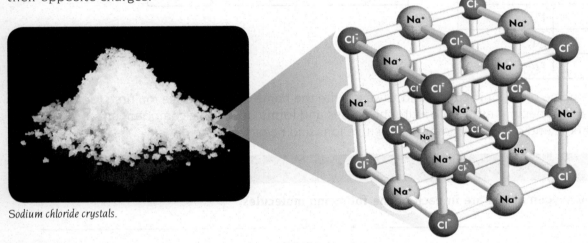

Sodium chloride crystals.

Ionic bonds are formed between metal atoms and non-metal atoms. In solution (dissolved in water), they can conduct electricity.

COVALENT BONDING

Covalent bonding occurs when two or more atoms share electrons. This most commonly occurs between atoms of non-metals. Because non-metal atoms 'want' to gain electrons, the elements involved will share electrons in an 'effort' to fill their outer shells.

A good example of a covalent bond is between two hydrogen atoms. Atoms of hydrogen (H) have only one electron in their outer shell. Since the capacity of this shell is two electrons, each hydrogen atom will 'want' to pick up a second electron. In an effort to pick up a second electron, hydrogen atoms will bond with nearby hydrogen atoms to form a hydrogen molecule.

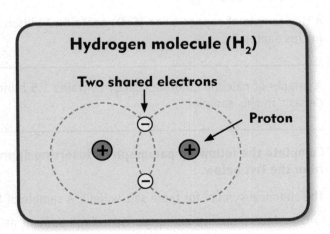

List of common ions

Positive ions	Symbol
Hydrogen	H^+
Sodium	Na^+
Potassium	K^+
Ammonium	NH_4^+
Silver	Ag^+
Lithium	Li^+
Magnesium	Mg^{2+}
Calcium	Ca^{2+}
Iron (II)	Fe^{2+}
Copper	Cu^{2+}
Zinc	Zn^{2+}
Lead	Pb^{2+}
Aluminium	Al^{3+}
Iron (III)	Fe^{3+}

Negative ions	Symbols
Chloride	Cl^-
Hydroxide	OH^-
Nitrate	NO_3^-
Hydrogen carbonate	HCO_3^-
Bromide	Br^-
Iodide	I^-
Oxide	O^{2-}
Sulfide	S^{2-}
Carbonate	CO_3^{2-}
Sulfate	SO_4^{2-}
Phosphate	PO_4^{3-}

When you write the name of an ionic compound, always write the name of the positive ion first, for example $Mg(NO_3)_2$ = magnesium nitrate. To write the names and formulae of ionic compounds you must know the symbols and charges of common positive ions and negative ions.

EXERCISES

A State how many oxygen atoms are in each of the following molecules.

1 CH_3COOH _____

2 $K_2Cr_2O_7$ _____

3 $PbNO_3$ _____

4 $Mg(OH)_2$ _____

5 $CuSO_4.5H_2O$ _____

6 $2KMnO_4$ _____

7 $CaCO_3$ _____

8 $2Ca(OH)_2$ _____

B Answer the following.

1 A tiny sample of copper oxide (CuO) contains 2 billion copper ions. What is the number of oxide ions present in this sample?

_____

2 A sample of calcium chloride ($CaCl_2$) contains 1.5 billion calcium ions. What is the number of chloride ions present in this sample?

_____

C Complete the following paragraph by inserting appropriate words in the blank space. Choose the words from the list below.

The chemical symbol for table salt is NaCl. A sample of NaCl contains equal numbers of _____ ions and _____ ions. Each sodium atom supplies one of its electrons to a chlorine atom. When a sodium atom loses one of its electrons it becomes a _____ ion. When a chlorine atom gains one electron from a sodium atom, it becomes a _____ ion. This means that sodium chloride is an _____ compound.

Words: positive, negative, electrons, ionic, sodium, chloride

ISBN: 978-0-17-018952-1

D Below are six models made by a student. Beside each model write its formula.

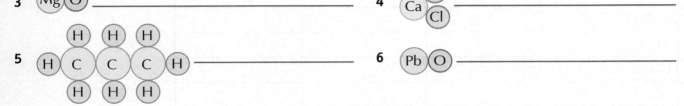

1 (H)(O)(H) _____

2 (H)(N)(O)(O)(O) _____

3 (Mg)(O) _____

4 (Ca)(Cl)(Cl) _____

5 (H)(H)(H)(H)(C)(C)(C)(H)(H)(H)(H) _____

6 (Pb)(O) _____

E Use the descriptions below to give the formulae of the compounds.

1 A compound containing 1 nitrogen atom and 2 oxygen atoms.

2 An aluminium compound containing 2 aluminium atoms and 3 oxygen atoms.

3 An oxide of carbon containing 1 carbon and 1 oxygen atom.

F Complete the chart below. The first row has been done for you.

Name of the compound	Formula	How many of each kind of atoms?	Total number of atoms
Copper carbonate	$CuCO_3$	1 copper atom 1 carbon atom 3 oxygen atoms	5 atoms
Copper hydroxide	$Cu(OH)_2$		
Sodium sulfate	Na_2SO_4		
Acetic acid	CH_3COOH		
Zinc sulfate	$ZnSO_4$		
Sodium thiosulfate	$Na_2S_2O_3$		
Urea	$CO(NH_2)_2$		
Lead nitrate	$Pb(NO_3)_2$		
Sodium bicarbonate (also known as sodium hydrogen carbonate)	$NaHCO_3$		
Zinc nitrate	$Zn(NO_3)_2$		

ISBN: 978-0-17-018952-1

G Complete the table below by writing the symbols of positive and negative ions present in each ionic compound. The first row has been done for you.

Formula of the compound	Positive ion	Negative ion
MgO	Mg^{2+}	O^{2-}
$CaCl_2$		
$CuSO_4$		
Al_2O_3		
$CaCO_3$		
HNO_3		
$Pb(NO_3)_2$		
$NaHCO_3$		

H Here are the formulae of 16 ionic compounds. Write the names of each.

1 $MgCl_2$ _____

2 Fe_2S_3 _____

3 KNO_3 _____

4 Na_2SO_4 _____

5 $(NH_4)_2CO_3$ _____

6 Al_2O_3 _____

7 $Mg(NO_3)_2$ _____

8 $Fe(OH)_3$ _____

9 $ZnSO_4$ _____

10 $KHCO_3$ _____

11 H_2S _____

12 $NaBr$ _____

13 $Fe(NO_3)_3$ _____

14 $CaCl_2$ _____

15 CuO _____

16 $ZnCl_2$ _____

ISBN: 978-0-17-018952-1

7 WRITING THE FORMULAE OF IONIC COMPOUNDS

If you know the symbols of the ions and their charges, you should be able to write the formula of almost all ionic compounds.

Ionic compounds contain positive ions and negative ions. The number of positive charges is balanced by an equal number of negative charges.

Here the magnesium ion has two positive charges and the oxide ion has two negative charges. They are equal so the formula for magnesium oxide is **MgO**, with one magnesium ion and one chloride ion.

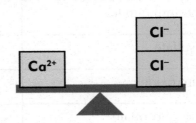

Here the calcium ion has two positive charges but the chloride ion has only one negative charge. The charges are not equal. Charges can be balanced and made equal by adding one more chloride ion.

Now there is one calcium ion and two chloride ions. So the formula for calcium chloride is **CaCl$_2$**. (If there is only one atom, like Ca in this case, we do not write a number 1 after the symbol.)

Using the 'cross and drop method' for working out a chemical formula

- Write down the name of the ionic compound.
- Write down the formula of the two ions involved. Always write the positive ion first.
- Cross and drop the numbers on the charges. (See example below)
- Do not put the sign of the charges on the final formula.
- If the ions have the same number of charges, you do not have to place any numbers on the final formula.
- If an ion has more than one atom and you want to place a number beside it, then you must put a bracket around the ion first.
- Some ions like carbonate ion CO_3^{2-} have a number in their formula. That number will stay with the ion all the time.

Example

Calcium hydroxide: Two ions involved are Ca^{2+} and OH^{1-}

Cross and drop the numbers on the charges.

The formula for calcium hydroxide is **Ca(OH)$_2$** (You do not have to write 1 beside Ca.)

ISBN: 978-0-17-018952-1

EXERCISES

A Complete the chart below.

Name of the compound	Positive ion	Negative ion	Formula
Magnesium chloride	Mg^{2+}	Cl^-	$MgCl_2$
Aluminium hydroxide			
Copper oxide			
Copper sulfate			
Calcium nitrate			
Ammonium chloride			
Calcium carbonate			
Iron (III) oxide			
Magnesium nitrate			
Zinc sulfate			
Sodium oxide			
Copper phosphate			

B Complete the chart by filling in the empty boxes.

Name of the compound	Positive ion	Negative ion	Formula
	Ca^{2+}		CaO
Calcium nitrate		NO_3^-	
	Li^+	CO_3^{2-}	
Aluminium sulfate			$Al_2(SO_4)_3$
	Fe^{3+}	O^{2-}	
	Ag^+		$AgCl$
Sodium carbonate		CO_3^{2-}	
			Na_2SO_4

C Write down the formula for each of these copper compounds.

1 Copper oxide _____

2 Copper chloride _____

3 Copper hydroxide _____

4 Copper carbonate _____

5 Copper sulfate _____

6 Copper sulfide _____

ISBN: 978-0-17-018952-1

D Magnesium hydroxide and sodium hydroxide are two ionic compounds. The formula for magnesium hydroxide is $Mg(OH)_2$ and the formula for sodium hydroxide is NaOH. Explain why the ratios of positive ions and negative ions are different in these two compounds.

In your answer you should consider:

- Charges on individual ions.
- Reason why ratio of positive ions to negative ions in $Mg(OH)_2$ is different from that of NaOH.

_____

_____

_____

_____

_____

_____

_____

_____

_____

_____

_____

_____

ISBN: 978-0-17-018952-1

When magnesium burns in air it makes a bright flame and produces a white powder called magnesium oxide. This is a chemical reaction. Instead of writing a long sentence to describe a chemical reaction, chemists use a chemical equation. You can write a word equation or a formula equation to sum up what happens in a chemical reaction.

magnesium + oxygen $\longrightarrow$ magnesium oxide (word equation)
$2Mg$ + O_2 $\longrightarrow$ $2MgO$ (formula equation)

A chemical equation has two parts, the **reactants** (left-hand side of the arrow) and the **products** (right-hand side of the arrow). In the above example, magnesium and oxygen are the reactants, and magnesium oxide is the product.

Example
When zinc metal is placed in dilute hydrochloric acid, zinc chloride and hydrogen gas are produced.

Reactants: zinc and hydrochloric acid
Products: zinc chloride and hydrogen
Word equation: zinc + hydrochloric acid $\longrightarrow$ zinc chloride + hydrogen
Formula equation: $Zn + 2HCl \longrightarrow ZnCl_2 + H_2$

BALANCING A CHEMICAL EQUATION

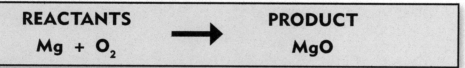

REACTANTS		PRODUCT
$Mg + O_2$	$\longrightarrow$	MgO

The above equation is not a balanced equation. In the reactant side it shows 2 oxygen atoms but in the product side there is only 1 oxygen atom. This is impossible because atoms are never 'lost' – they just change partners.

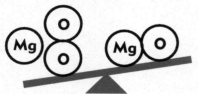

$Mg + O_2 \rightarrow MgO$ ✗

In a balanced chemical equation, the number of atoms of each kind after the reaction is exactly the same as the number of atoms before the reaction.

You can balance this by adding another magnesium atom on the reactant side and a magnesium oxide molecule on the product side as shown below.

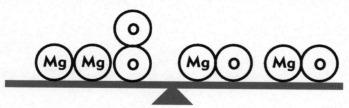

✓

$2Mg + O_2 \rightarrow 2MgO$

Elements like hydrogen, oxygen, nitrogen, chlorine, fluorine, iodine and bromine exist in nature as **diatomic** molecules so write their formulae as H_2, O_2, N_2, Cl_2, F_2, I_2, Br_2.

ISBN: 978-0-17-018952-1

Easy steps to balance an equation

1 Write down the word equation:

Iron + Oxygen ⟶ Iron (III) oxide

2 Write down the formula equation: **Fe + O$_2$ ⟶ Fe$_2$O$_3$**

3 Count the number of atoms of the first element. In this example it is Fe, on both sides of the equation. There is only one Fe atom on the reactant side and two Fe atoms on the product side. If one side requires more, you add more of that element on that side. So you add one more Fe on the reactant side.

Fe + O$_2$ ⟶ Fe$_2$O$_3$
Fe

Now we have two Fe atoms on both sides of the equation.

4 Balance the second element in the same way. There are two oxygen atoms on the reactant side and three oxygen atoms on the product side. The only way to balance the oxygen atom is to add two molecules of oxygen in the reactant side and another molecule of Fe$_2$O$_3$ in the product side.

Fe + O$_2$ ⟶ Fe$_2$O$_3$
Fe   O$_2$ Fe$_2$O$_3$
**   O$_2$**

Now we have six oxygen atoms on both sides of the equation.

5 Go back to the first element and check its number on both sides of the equation. Add more if needed.
There are two Fe atoms on the reactant side and four Fe atoms on the product side. This can be balanced by adding two more Fe atoms on the reactant side.

Fe + O$_2$ ⟶ Fe$_2$O$_3$
Fe   O$_2$ Fe$_2$O$_3$
Fe   O$_2$
Fe

Now we have four Fe atoms and six oxygen atoms on both sides of the equation.

6 When all atoms are balanced, rewrite the equation showing the number of atoms or molecules in front of the symbol or the formula on both sides of the equation. There are four Fe atoms and three oxygen molecules (six oxygen atoms) on the reactant side and two molecules of Fe$_2$O$_3$ on the product side. Two molecules of Fe$_2$O$_3$ has four Fe atoms and six oxygen atoms. So you can write the balanced equation as shown below.

4Fe + 3O$_2$ ⟶ 2Fe$_2$O$_3$

EXERCISES

A   Write down <u>word</u> equations for the following chemical reactions.

1 Calcium reacts with oxygen in the air to form calcium oxide.

_____ + _____ ⟶ _____

2 When lithium reacts with water, it makes lithium hydroxide and hydrogen gas.

_____ + _____ ⟶ _____ + _____

ISBN: 978-0-17-018952-1

3 Sodium hydroxide reacts with hydrochloric acid to form sodium chloride and water.

_____ + _____ ⟶ _____ + _____

4 Marble chips (calcium carbonate) react with hydrochloric acid to form calcium chloride, water and carbon dioxide gas.

_____ + _____ ⟶ _____ + _____ + _____

5 Methane (CH_4) gas burns in oxygen to form carbon dioxide and water.

_____ + _____ ⟶ _____ + _____

B Write down <u>balanced</u> chemical equations for the following word equations.

1 calcium + oxygen ⟶ calcium oxide

_____ + _____ ⟶ _____

2 magnesium + water ⟶ magnesium hydroxide + hydrogen

_____ + _____ ⟶ _____ + _____

3 magnesium + sulfuric acid ⟶ magnesium sulfate + hydrogen

_____ + H_2SO_4 ⟶ + _____ + _____

4 zinc + oxygen ⟶ zinc oxide

_____ + _____ ⟶ _____

C Balance the following chemical equations. Rewrite the whole equation in the space below.

1 $H_2 + O_2 \longrightarrow H_2O$

_____

2 $Na + H_2O \longrightarrow NaOH + H_2$

_____

3 $Ca(OH)_2 + HCl \longrightarrow CaCl_2 + H_2O$

_____

4 $Fe + Cl_2 \longrightarrow FeCl_3$

_____

5 $K_2O + HCl \longrightarrow KCl + H_2O$

_____

6 $Na_2CO_3 + HCl \longrightarrow NaCl + CO_2 + H_2O$

_____

7 $Mg + HCl \longrightarrow MgCl_2 + H_2$

_____

8 $CuCO_3 + H_2SO_4 \longrightarrow CuSO_4 + CO_2 + H_2O$

_____

9 $CaO + HCl \longrightarrow CaCl_2 + H_2O$

_____

ISBN: 978-0-17-018952-1

PARTICLE THEORY

The kinetic theory of matter, which is also known as the particle theory of matter, says that matter consists of very small particles which are constantly moving. Particles can mean atoms or molecules or ions. The speed at which these particles move depends on the amount of energy the particles have and the state of matter. In solids and liquids, the space between the particles is very small compared to the space between particles in gases. The attractive forces between the particles are also very strong in solids. When heated, particles move faster and have more frequent collisions, therefore taking up more space and making an object expand. When cooled, particles move more slowly, having less frequent collisions, therefore taking up less space making an object contract.

Physical change.

PHYSICAL AND CHEMICAL CHANGES

If liquid water is boiled to produce steam, it is still water. Likewise frozen water (ice) is still water. Melting, boiling, or freezing are examples of **physical changes**, because they do not affect the internal composition of the items involved. It only changes the state of matter. These processes can be reversed easily by adding or removing heat.

A **chemical change**, on the other hand, occurs when the chemical composition changes – one substance is transformed into another. Water can be chemically changed, for instance, when an electric current is run through it, separating it into oxygen and hydrogen gas. When wood burns it becomes ash and CO_2. Obviously, the ashes cannot be simply frozen to turn them back into wood again. Some of the characteristics by which you can recognise chemical changes:

Physical change.

- Energy in the form of heat is either produced or absorbed. It takes energy to break chemical bonds.
- Bubbles of gas may appear.
- A precipitate may form. A precipitate is an insoluble solid product formed when two soluble substances are mixed together.
- Light may be emitted. Most combustion reaction produces light.
- A colour change may occur.
- A change in smell or taste may occur.
- A change in the melting point and boiling point may occur.
- The change is not easily reversed.

COLLISION THEORY TO EXPLAIN CHEMICAL REACTIONS

Two substances can only react together if they come into contact with each other. Their particles first have to collide. Not all collisions will result in a chemical reaction. The particles need to collide with enough energy, and also, the particles need to collide the right way around (orientation).

Chemical change.

ISBN: 978-0-17-018952-1

The energy of collision

The minimum amount of energy required before a reaction can occur is called **activation energy**. If the particles collide with less energy than the activation energy, nothing important happens. Only those collisions which have energies equal to or greater than the activation energy result in a reaction. Any chemical reaction results in the breaking of some bonds and the making of new ones. Obviously some bonds have to be broken before new ones can be made. Activation energy is needed to break some of the original bonds. If the collisions are gentle, the bond breaking process will not happen.

Orientation of particles during collision

A reaction can occur only if the reactant particles are positioned correctly during collision. For example, an organic compound called ethane (formula $Ch_3=CH_3$ or C_2H_6) reacts with hydrochloric acid (HCl) to form a compound called chloroethane (formula $CH_3 - CH_2 - Cl$ or C_2H_5Cl). For this reaction to occur, the molecules of ethane and hydrochloric acid must collide in a certain way. See the illustration below.

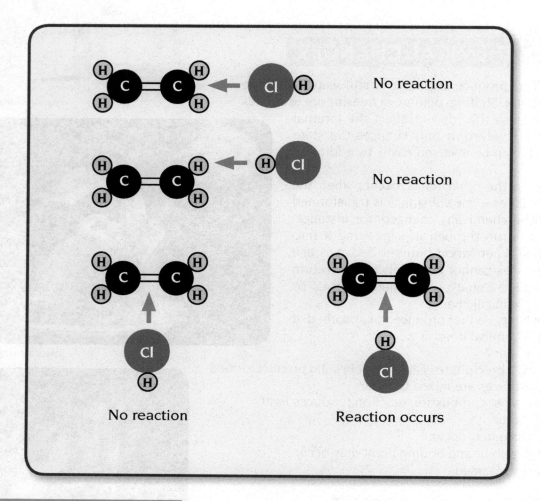

No reaction

No reaction

No reaction   Reaction occurs

REACTION RATES

'Rate of reaction' means *'the speed of the reaction'* or *'how fast the reaction happens'*. The rate of a reaction can be measured as the 'rate of formation of product' or the 'rate of removal of reactant'. This can be explained simply as how fast the products are formed or how fast the reactants are used up. Rate of a chemical reaction that produces a gas can be measured by collecting the gas and measuring its volume as it is produced. For example, the reaction between calcium carbonate and hydrochloric acid that produces carbon dioxide gas.

ISBN: 978-0-17-018952-1

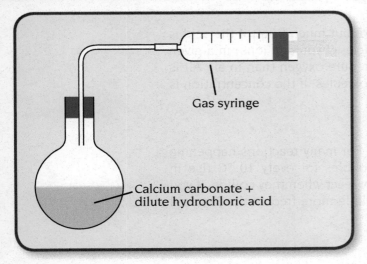

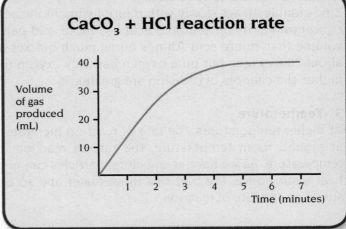

Some reactions (for example the rusting of iron) occur very slowly. Others (for example burning of magnesium ribbon) occur very quickly. The rate of a chemical reaction depends on several factors.

Factors affecting the rate of a chemical reaction are:

1 Surface area of the reactants

The more finely divided the solid is, the faster the reaction happens. A powdered solid will normally produce a faster reaction than if the same mass is present as a single lump. The powdered solid has a greater surface area than the single lump. More surface area means more effective collision of particles.

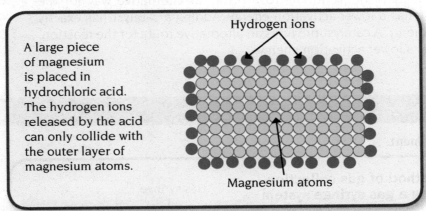

A large piece of magnesium is placed in hydrochloric acid. The hydrogen ions released by the acid can only collide with the outer layer of magnesium atoms.

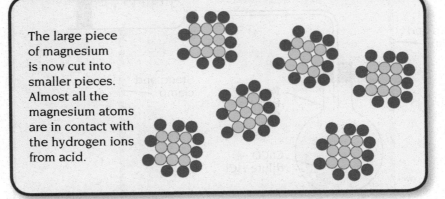

The large piece of magnesium is now cut into smaller pieces. Almost all the magnesium atoms are in contact with the hydrogen ions from acid.

ISBN: 978-0-17-018952-1

2 Concentration of reactants

Zinc granules react slowly with dilute hydrochloric acid, but much faster if the acid is concentrated. A concentrated acid has more acid particles (hydrogen ions) in a given volume than dilute acid. Things burns much quicker in pure oxygen than in air. Air is about 20% oxygen but pure oxygen has only oxygen molecules. If the concentration is higher, the chances of collision are greater.

3 Temperature

At higher temperatures, the rate of reaction increases. For many reactions happening at around room temperature, the rate of reaction doubles for every 10 °C rise in temperature. As we have seen before, particles can only react when they collide. If you heat a substance, the particles move faster and so collide more frequently. That will speed up the rate of reaction.

4 Pressure

Increasing or decreasing pressure may not have any effect on the rate of reaction involving solids and liquids, but it does affect the rate of reaction between gases. Increasing the pressure of a gas is like increasing its concentration. The way to increase gas pressure is to squeeze it into a smaller volume. The same mass in a smaller volume means that its concentration is higher.

5 Catalysts

A catalyst is a substance which speeds up a reaction, but is chemically unchanged at the end of the reaction. When the reaction has finished there is exactly the same mass of catalyst as at the beginning.

To increase the rate of a reaction you need to increase the number of successful collisions. One possible way of doing this is to provide an alternative way for the reaction to happen which has a lower activation energy. Adding a catalyst has exactly this effect on activation energy. A catalyst provides an alternative route for the reaction. That alternative route has a lower activation energy.

EXERCISES

A Gas collection experiment

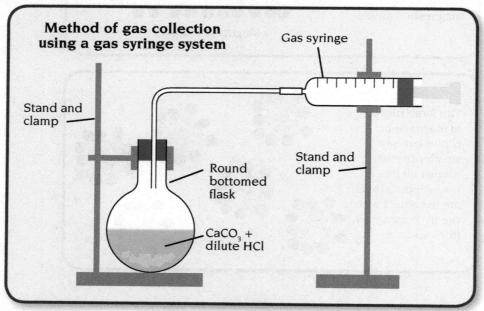

Method of gas collection using a gas syringe system

Gas syringe

Stand and clamp

Round bottomed flask

Stand and clamp

$CaCO_3$ + dilute HCl

Rowena did an experiment as shown in the diagram. She placed calcium carbonate and dilute hydrochloric acid (0.5 mol) in a conical flask, and collected the gas produced in a gas syringe. She recorded the volume of gas produced in every half minute. These are **Data 1**, on the next page.

ISBN: 978-0-17-018952-1

Data 1

Time (minutes)	Volume of gas (mL)
0.0	0
0.5	5
1.0	11
1.5	17
2.0	23
2.5	27
3.0	30
3.5	32
4.0	36
4.5	39
5.0	40
5.5	40
6.0	40
6.5	40
7.0	40

She then used the same equipment and conducted a second experiment. She took the same mass of calcium carbonate and the same volume of 0.1 mol hydrochloric acid. Given below are the data she has recorded. These are **Data 2** given below.

Data 2

Time (minutes)	Volume of gas (mL)
0.0	0
0.5	8
1.0	15
1.5	20
2.0	25
2.5	30
3.0	38
3.5	40
4.0	40
4.5	40
5.0	40
5.5	40
6.0	40
6.5	40
7.0	40

1 On the grid below plot two separate line graphs for Data 1 and Data 2. Label the axes and provide a key to identify the two graphs. Give your graph a title.

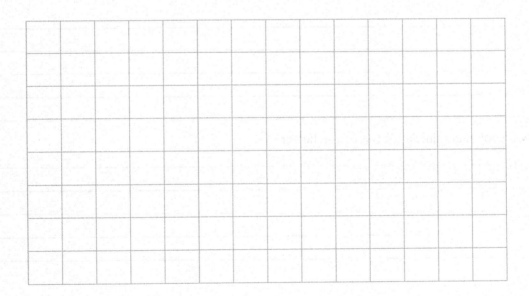

2 Explain why each graph levels off.

_____
_____
_____

3 Explain a reason for any difference between the two graphs.

_____
_____
_____

ISBN: 978-0-17-018952-1

4 Explain two other ways Rowena could increase the rate of this chemical reaction without changing the mass and volume of the reactants.

_____

_____

_____

_____

_____

B **Give scientific explanations for the following:**

1 A mixture of gases reacts faster when the volume it occupies is decreased.

_____

_____

_____

_____

2 Iron filings rust faster than an iron nail.

_____

_____

_____

3 There is a danger of explosions in places such as coal mines where there are large quantities of powdered coal dust.

_____

_____

_____

_____

_____

4 Pikelets will cook more quickly if the pan is hotter.

_____

_____

_____

_____

_____

C

1 Food is spoiled mainly by the action of bacteria and fungi. These micro-organisms secrete enzymes to break down food. One practical measure to slow down food spoilage is to refrigerate the food. Use your knowledge of factors affecting the rate of chemical reaction to explain how refrigeration slows down food spoilage.

_____

_____

_____

_____

_____

ISBN: 978-0-17-018952-1

2 Another way to preserve food is to vacuum-pack it. Explain how vacuum-packing keeps food safe much longer than just keeping it in closed containers.

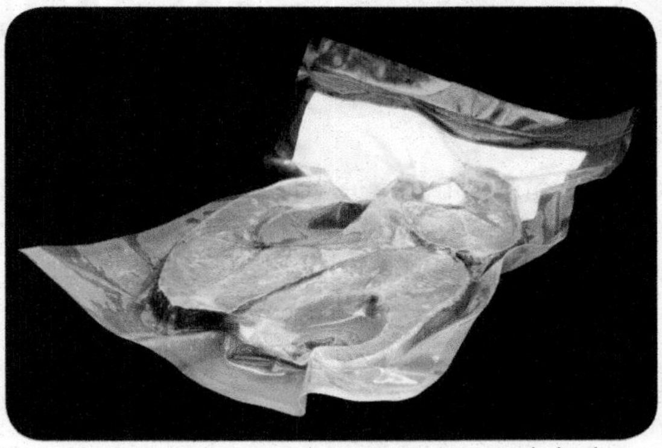

Vacuum-packed salmon.

_____

_____

_____

_____

_____

_____

_____

_____

_____

_____

_____

_____

3 Kerry has an open fireplace at home. When lighting a fire she notices that thin kindling wood is much easier to light than large logs. Use your knowledge of particle theory to explain this.

Logs.

Kindling.

_____

_____

_____

_____

_____

4 Kerry has also learnt from her outdoor education class that blowing air on a fire increases the rate of combustion. Use your knowledge of factors affecting the rate of a reaction to explain this.

_____

_____

_____

_____

_____

_____

_____

_____

ACIDS

Acids are common in everyday life. Some are weak while others are strong. Many are dangerous to touch or taste. Most weak acids found in nature are quite safe to taste. They are called **organic** acids. See the list below.

Citrus fruit, e.g. oranges, lemons		Citric acid
Grapes		Tartaric acid
Apples		Malic acid
Yoghurt		Lactic acid
Rhubarb leaves (Warning: poisonous)		Oxalic acid
Vinegar		Acetic acid
Ants		Formic acid

Strong acids are corrosive and can eat away materials like metals and fabrics. Examples of strong **mineral** acids found in a school laboratory are: hydrochloric acid, HCl; sulfuric acid, H_2SO_4; nitric acid, HNO_3.

Strong acids can be diluted by adding water. Most fizzy drinks contain dissolved carbon dioxide. When carbon dioxide dissolves in water it forms an acid called **carbonic acid** (H_2CO_3).

All non-metal oxides are acidic, which means they can dissolve in water to form acidic solutions. **Sulfur dioxide** (SO_2) dissolves in water to form **sulfurous acid** (H_2SO_3), and **sulfur trioxide** dissolves in water to form sulfuric acid (H_2SO_4).

As you have seen, there are many different acids. We have also noticed that all of them contain hydrogen (HCl, H_2SO_4, HNO_3). They are soluble in water and can conduct electricity. It shows that they all form ions in water. Hydrochloric acid (HCl) dissociates into H^+ ions and Cl^- ions; sulfuric acid H_2SO_4 dissociates into H^+ ions and SO_4^{2-} ions and so on.

Properties of acids

- Acids (weak acids) have a sour taste. (BUT DO NOT TASTE THEM.)
- Acids change the colour of **indicators** such as **litmus**.
- Acids react with certain metals to form hydrogen gas.
- Acids react with carbonates and bicarbonates to form carbon dioxide gas.
- Acids react with bases or **alkalis** to form some kind of 'salt' and water.
- Acids provide H^+ ions.

BASES AND ALKALIS

Bases are another group of everyday chemicals. They have special properties which are quite different from those of acids. Often they behave in opposite ways to acids.

Acids change the colour of blue litmus to red while bases or alkalis change red litmus into blue. Soluble bases are called **alkalis**.

Metal oxides and hydroxides are examples of bases. Soluble metal hydroxides like sodium

ISBN: 978-0-17-018952-1

hydroxide, NaOH; potassium hydroxide, KOH; ammonium hydroxide, NH_4OH; and calcium hydroxide, $Ca(OH)_2$ are the common alkalis found in a school laboratory.

Most household cleaning products contain bases. Oven cleaners contain NaOH (sodium hydroxide), which is commonly called **caustic soda**. Soap is produced when an alkali reacts with fat or oil. This explains why alkalis feel soapy to touch: they react with oils on your skin.

Toothpaste contains a weak base to neutralise the acid produced by mouth bacteria. Antacid tablets used for treating indigestion also contain basic compounds to neutralise excess acid in the stomach.

Properties of bases or alkalis

- Weak bases have a bitter taste and feel slippery to touch. (NEVER TASTE OR TOUCH A STRONG BASE.)
- Bases change the colour of indicators. They reverse the change produced by acids. (See the next chapter).
- Bases react with acids to form salt and water.
- Soluble bases release hydroxide ions (OH^-) during chemical reactions.

EXERCISES

A   Answer the following.

1 Name four weak organic acids.
_____
_____
_____
_____

2 Write down the names and formulae of three mineral acids found in a school laboratory.
_____
_____
_____

3 Write down the names and formulae of two acidic oxides.
_____
_____
_____

4 You are given two liquids in unlabelled bottles. One bottle contains pure distilled water and the other bottle contains dilute hydrochloric acid. Explain three different tests you could use to distinguish acid from water. (Tasting is not allowed.)
_____
_____
_____

5 Write down the names and formulae of four common alkalis found in a school laboratory.
_____
_____
_____

6 State one chemical difference between a base and an alkali.
_____
_____

7 Explain why most toothpastes contain a weak base.
_____
_____

8 Explain why alkalis feel slippery to touch.
_____
_____

ISBN: 978-0-17-018952-1

9 Explain why indigestion tablets and powders relieve acid stomach.

_____

_____

10 The Australian box jellyfish inject a base when they sting. Suggest what could be the best first-aid for relieving the pain caused by their sting. Explain how your suggestion works.

_____

_____

_____

_____

11 Name the two substances that always form when an acid reacts with a base.

_____

_____

B The table below compares the properties of two compounds, sodium hydroxide and hydrochloric acid. Place a ✓ or × in the appropriate boxes.

Formula of compound	Contains H^+ ion	Contains OH^- ion	Releases hydrogen when reacts with acid	Changes the colour of blue litmus	Changes the colour of red litmus	An acid	A base or an alkali
NaOH							
HCl							

C Name the following.

1 Gas produced when metals react with acid. _____

2 Gas produced when marble chips react with acid. _____

3 Acid found in citrus fruit such as lemons and oranges. _____

4 Acid found in household vinegar. _____

5 Ions released by all acids. _____

6 Ions released by all alkalis. _____

7 An indicator used for testing acids. _____

D Write T (true) or F (false) beside each of the statements below.

☐ **1** Weak acids taste sour.

☐ **2** Most metals react with dilute acids to form salt and hydrogen gas.

☐ **3** Acids turn red litmus into blue.

☐ **4** Your body makes acid.

☐ **5** Fizzy drinks change the colour of blue litmus into red.

☐ **6** All sour liquids contain alkalis.

☐ **7** Toothpaste and oven cleaners contain bases.

☐ **8** All bases are soluble in water.

☐ **9** Hydroxide ions are negatively charged particles.

ISBN: 978-0-17-018952-1

E Match the common name of the alkali with its chemical name and formula. You may draw straight lines to join these boxes.

Common name	Chemical name	Formula
Caustic soda	Magnesium hydroxide	KOH
Caustic potash	Ammonium hydroxide	$Mg(OH)_2$
Limewater	Sodium hydroxide	NH_4OH
Milk of magnesia	Potassium hydroxide	$Ca(OH)_2$
Ammonia	Calcium hydroxide	NaOH

F Match the two lists below.

1 Acetylsalicylic acid   **A** Acid produced by certain ants.
2 Citric acid   **B** Acid produced in the stomach of humans.
3 Malic acid   **C** Acid found in vinegar.
4 Tartaric acid   **D** Acid found in aspirin.
5 Formic acid   **E** Acid found in oranges and lemons.
6 Lactic acid   **F** A strong mineral acid.
7 Oxalic acid   **G** Acid found in apples.
8 Hydrochloric acid   **H** Acid found in grapes.
9 Sulfuric acid   **I** Acid found in yoghurt.
10 Acetic acid   **J** Acid found in plants like rhubarb.

Answers

1	2	3	4	5	6	7	8	9	10

G Complete the table below.

Key: H = ●   S = ●   N = ●   Cl = ●   O = ●

Simplified molecular diagram	Molecular formula	Name of acid	Positive ion	Negative ion	Salt produced by the acid
		Sulfuric acid			Sulfate
	HNO_3			NO_3^-	
			H^+		

ISBN: 978-0-17-018952-1

INDICATORS

An indicator can be used to tell whether a substance is acidic, basic or neutral. Most indicators are dyes extracted from plants. They change their colour when an acid or a base is added. Some common acid-base indicators found in the laboratory are listed below.

Indicator	Colour in neutral solution	Colour in acidic solution	Colour in basic solution
Phenolphthalein	Colourless	Colourless	Pink
Litmus	Blue or red	Red	Blue
Methyl orange	Orange	Pink	Orange
Bromothymol blue	Blue	Yellow	Blue
Universal indicator	Green	Shades of red	Blue

pH SCALE

This is a scale from 0 to 14 used for indicating the strength of an acid or a base. Any solution with a pH value below 7 is an acid. This is a measure of the H^+ ion concentration of the solution. The higher the H^+ ion concentration, the smaller the pH value. A solution with a pH value 2 is a much stronger acid than a solution with a pH value 6. A neutral solution (neither acidic nor basic) has a pH value of 7. Basic substances have pH values higher than 7. Remember that soluble bases are called alkalis.

The pH scale

0	1	2	3	4	5	6	7	8	9	10	11	12	13	14

ACID ←   NEUTRAL   → BASE OR ALKALI

STRONG   WEAK   WEAK   STRONG

NEUTRALISATION REACTION

Any reaction between an acid and a base is called a neutralisation reaction. When an acid reacts with a base or an alkali, the products formed are a salt and water.

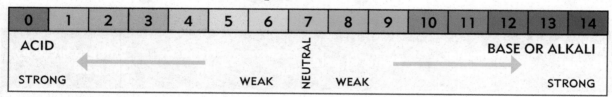

$$\text{ACID} + \text{BASE} \longrightarrow \text{SALT} + \text{WATER}$$

Hydrochloric acid always produces a chloride salt. For example:

hydrochloric acid + sodium hydroxide ⟶ sodium *chloride* + water
    HCl   +   NaOH   ⟶   NaCl   + H_2O

Sulfuric acid always gives a sulfate salt. For example:

sulfuric acid + copper oxide ⟶ copper *sulfate* + water
  H_2SO_4 +   CuO   ⟶ $CuSO_4$ + H_2O

Nitric acid produces nitrates, carbonic acid produces carbonates and acetic acid produces acetate salts.

ISBN: 978-0-17-018952-1

Practical applications of neutralisation

The odour of 'not fresh' fish is due to bases called amines. Adding lemon juice can reduce this odour by neutralising these bases.

Indigestion, or 'heart burn', is due to the build-up of acid in the stomach. This can be cured by taking antacid tablets or powders. These remedies contain weak bases such as sodium bicarbonate, magnesium hydroxide and aluminium hydroxide.

EXERCISES

A Answer the following.

1 Describe what is meant by a neutralisation reaction.

_____

_____

_____

_____

2 State what acid-base indicators are used for.

_____

_____

3 Name four acid-base indicators and describe their colour change in acids, in bases and in neutral solutions.

Name of indicator	Colour in neutral solution	Colour in acidic solution	Colour in basic solution

B Give scientific reasons for the following.

1 The smell of 'not-fresh' fish can be removed by adding vinegar or lemon juice.

_____

_____

2 Stings of box jellyfish can be treated with vinegar. (Note this does NOT work for stings from all kinds of jellyfish.)

_____

_____

_____

ISBN: 978-0-17-018952-1

C The table below shows the pH value of certain substances.

1 Complete the table by writing the description as '**strongly acidic**', '**strongly basic**', '**moderately acidic**', '**moderately basic**', '**slightly acidic**', '**slightly basic**' or '**neutral**' in the last column.

Substance	pH value	Description
Soft drink	4.5	
Distilled water	7	
Gastric juice in the stomach	2.4	
Dishwashing liquid	13	
Concentrated hydrochloric acid	1	
Concentrated sodium hydroxide	14	
Polluted water	9	
Rainwater	6.5	
Human blood	7.3	
Human urine	6.1	

2 Name the acid responsible for making

   a the soft drinks acidic. _____

   b gastric juice acidic. _____

3 State what colour litmus would be in

   a dishwashing liquid. _____

   b rainwater. _____

4 Name two substances from the list above which, on mixing together, produce sodium chloride.

_____ and _____

D Ryan did an experiment as shown below. Study this to answer the following questions.

1

2

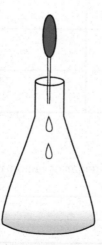

3

4

He put 20 mL of sodium hydroxide solution into a flask using a pipette.	He then added two drops of phenolphthalein into the flask using a dropper. The solution turned pink.	From a burette he added dilute HCl into the flask. At one point, the pink colour of the solution cleared and he stopped adding the acid.	He transferred the solution into an evaporating dish and heated it above a Bunsen flame.

ISBN: 978-0-17-018952-1

1 State the purpose of adding phenolphthalein to the flask.

_____

_____

2 Explain why the colour of the solution cleared at stage 3.

_____

_____

_____

3 Name two substances present in the solution at stage 3.

_____

_____

4 State the purpose of heating the solution at stage 4.

_____

_____

_____

E Answer the following.

1 What is the pH scale used to measure?

_____

_____

2 Suggest a pH value for toothpaste and then discuss why this pH is probably the best.

_____

_____

_____

_____

3 You are given a soluble tablet. State how you would find out if this tablet is acidic, basic or neutral.

_____

_____

_____

4 After taking an antacid tablet to soothe heartburn, a person burps. Explain the chemistry behind this.

_____

_____

_____

F Hannah did an experiment to study the reaction between magnesium oxide and dilute hydrochloric acid and wrote the following chemical equation to represent the reaction.

$$MgO + HCl \longrightarrow MgCl + H_2$$

This equation contains THREE errors.

1 Identify THREE errors in this equation.

_____

_____

_____

2 Rewrite a correct balanced equation for the reaction between magnesium oxide and hydrochloric acid.

_____

_____

ISBN: 978-0-17-018952-1

G In her chemistry class, Hannah did another experiment to neutralise hydrochloric acid by adding sodium hydroxide to it. She has learnt in her science class that adding sodium hydroxide to hydrochloric acid affects the pH of the solution.

1 Discuss how she could have determined when the hydrochloric acid has been completely neutralised by sodium hydroxide and how the pH of the solution changes during this reaction.

_____

_____

_____

_____

_____

_____

_____

2 Discuss the similarities and differences between the reactions of:

• Magnesium and hydrochloric acid

• Magnesium oxide and hydrochloric acid.

Support your answers with both word equations and balanced chemical equations.

_____

_____

_____

_____

_____

_____

_____

_____

_____

_____

_____

_____

_____

_____

_____

_____

H Answer the following.

1 Complete the chart below by writing the colour you would observe

Substance	Colour with blue litmus	Colour with universal indicator
Dilute hydrochloric acid		
Distilled water		
Limewater		

ISBN: 978-0-17-018952-1

2 Use the information below to answer the questions that follow.

pH	Example
0	HCl
1	Stomach acid
2	Lemon juice
3	Vinegar
4	Orange juice
5	Polluted rainwater
6	Milk
7	Pure water
8	Egg white
9	Toothpaste with baking soda
10	Antacid
11	Ammonia
12	Limewater
13	Oven cleaner
14	Sodium hydroxide

State the most strongly basic and acidic substance listed in the chart.

Most strongly basic: _____

_____

Most strongly acidic: _____

_____

3 Which fruit juice is more acidic, lemon juice or orange juice? Give a reason for your choice.

_____

_____

_____

4 Explain why toothpaste needs to have such a high pH. You must consider the following when you write the answer.

- Conversion of sugar into acid by bacteria.
- Neutralisation of acid.

_____

_____

_____

_____

_____

5 Antacids are medicines used to relieve indigestion, upset stomach and heartburn. They contain ingredients such as aluminium hydroxide, calcium carbonate, magnesium hydroxide and sodium bicarbonate. Use this information and the information from the chart above to discuss how antacids work to relieve problems such as indigestion, upset stomach and heartburn.

_____

_____

_____

_____

_____

_____

_____

_____

ISBN: 978-0-17-018952-1

6 Some metals react with water to form metal hydroxides. Andy places a piece of magnesium ribbon in a beaker of hot water. He saw gas bubbles appear at the edges of magnesium ribbon. He then tested the solution with a red litmus paper and the paper turns blue.

Discuss the reason for this colour change. Write a chemical equation to explain what had happened.

_____

_____

_____

_____

_____

7 A teacher gives two white powders to David. One powder is calcium oxide and the other powder is calcium carbonate. The only other chemical available to the student is dilute hydrochloric acid.

Explain how David will identify which powder is calcium oxide and which powder is calcium carbonate and describe what will happen. Write chemical equations to support your answer.

_____

_____

_____

_____

_____

_____

8 Lucy put 10 mL of dilute hydrochloric acid in a conical flask. She then added five drops of universal indicator. The solution turned red as she expected. She then added sodium hydroxide solution drop by drop into the flask from a burette. She noticed a series of colour changes until the solution turned purple.

In the boxes provided, fill in the colours that Lucy probably saw during this experiment.

Red → ☐ → ☐ → ☐ → Purple

9 Lucy says at one stage, the acid in the flask is completely neutralised by sodium hydroxide. Describe what she would observe when this happens.

_____

_____

_____

_____

_____

10 Adding sodium hydroxide drop by drop into hydrochloric acid changes the pH value of the solution. Describe the effect of adding sodium hydroxide to hydrochloric acid on the pH of the solution.

_____

_____

_____

_____

_____

ISBN: 978-0-17-018952-1

I   A model of a volcano

Zoe made a model volcano by adding vinegar (acetic acid) to baking powder (sodium hydrogen carbonate). The diagram shows her set-up.

Cross section of Zoe's volcano

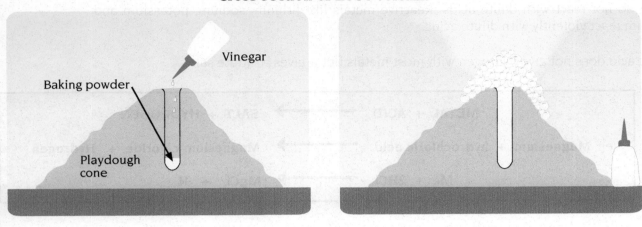

1 Write a word equation for the reaction between sodium hydrogen carbonate and acetic acid.

_____

_____

_____

2 Describe an observation that Zoe would make during her experiment.

_____

_____

_____

_____

3 With the help of your word equation for question 1, explain what happened.

_____

_____

_____

_____

_____

_____

ISBN: 978-0-17-018952-1

Most metals react with dilute acids to form salt and hydrogen gas. Metals like copper and silver do not react with dilute acids. Reactive metals like lithium, sodium, potassium and calcium react violently with dilute acids.

Nitric acid does not give hydrogen with most metals but it gives a nitrate salt .

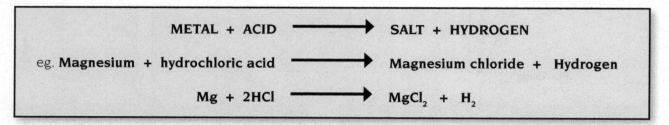

METAL + ACID $\longrightarrow$ SALT + HYDROGEN

eg. **Magnesium + hydrochloric acid** $\longrightarrow$ **Magnesium chloride + Hydrogen**

$$Mg + 2HCl \longrightarrow MgCl_2 + H_2$$

Hydrogen gas can be prepared in the laboratory by placing a metal, usually zinc, in dilute acid like hydrochloric acid. The diagram below shows how this can be done.

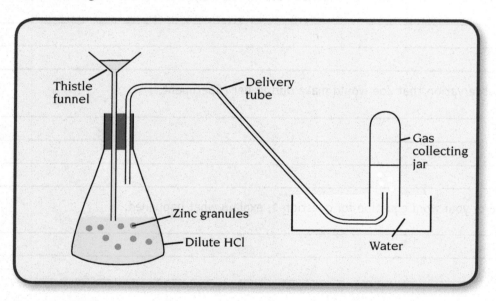

Hydrogen is a colourless and odourless gas. It is lighter than air and is insoluble in water. Because of these properties, hydrogen is collected by the **downward displacement** of water (see the diagram). A hydrogen-air mixture explodes near a flame and makes a squeaky pop sound.

METAL OXIDE AND ACIDS

All metal oxides are **bases**. They react with acids to form salt and water.

METAL OXIDE + ACID $\longrightarrow$ SALT + WATER

eg. **Magnesium oxide + Hydrochloric acid** $\longrightarrow$ **magnesium chloride + Water**

$$MgO + 2HCl \longrightarrow MgCl_2 + H_2O$$

ISBN: 978-0-17-018952-1

All metal hydroxides, like metal oxides, are bases. Soluble metal hydroxides are called **alkalis**. They react with acids to form salt and water.

METAL HYDROXIDE + ACID ⟶ SALT + WATER

eg. Calcium hydroxide + Hydrochloric acid ⟶ Calcium chloride + Water

Some metal hydroxides are **amphoteric**. Such metal hydroxides can form salt with acids and bases. Examples of amphoteric hydroxides are zinc hydroxide, aluminium hydroxide and lead hydroxide.

METAL CARBONATES AND HYDROGEN CARBONATES WITH ACID

Metal carbonates and hydrogen carbonates react with acid to form a salt, water and carbon dioxide gas.

METAL CARBONATE + ACID ⟶ SALT + WATER + CARBON DIOXIDE

eg. Calcium carbonate + Hydrochloric acid ⟶ Calcium chloride + Water + Carbon dioxide

$$CaCO_3 + 2HCl \longrightarrow CaCl_2 + H_2O + CO_2$$

Carbon dioxide gas can be made in the laboratory by the reaction of metal carbonates such as marble chips (calcium carbonate) and acid. Carbon dioxide can be tested by bubbling the gas through **limewater**. Limewater is a dilute solution of $Ca(OH)_2$: calcium hydroxide. It goes cloudy because of the formation of white insoluble calcium carbonate. Most metal carbonates are insoluble in water. The only exceptions are sodium, potassium and ammonium carbonates.

$$Ca(OH)_2 + CO_2 \longrightarrow CaCO_3 + H_2O$$

Limewater   White insoluble calcium carbonate

Most metal carbonates decompose when heated to form metal oxides and carbon dioxide gas.

METAL CARBONATE ⟶ METAL OXIDE + CARBON DIOXIDE

eg. Calcium carbonate ⟶ Calcium oxide + Carbon dioxide

$$CaCO_3 \longrightarrow CaO + CO_2$$

Baking powder contains sodium hydrogen carbonate ($NaHCO_3$) and an acidic compound like cream of tartar. When baking powder is added to moist flour while baking, it releases carbon dioxide gas which makes the dough rise.

Sodium hydrogen carbonate + Hydrochloric acid ⟶ Sodium chloride + Carbon dioxide + Water

$$NaHCO_3 + HCl \longrightarrow NaCl + CO_2 + H_2O$$

ISBN: 978-0-17-018952-1

METAL OXIDES AND HYDROXIDES

Metal oxides and hydroxides are **basic** substances.

Metal oxides are formed when metals react with oxygen. They can also be prepared by heating certain metal carbonates.

> **Calcium + Oxygen ⟶ Calcium oxide**
>
> **Calcium carbonate ⟶ Calcium oxide + Carbon dioxide**

Most metal oxides are insoluble in water. Soluble metal oxides react with water to form metal hydroxides. Such metal hydroxides are called **alkalis**.

> **Calcium hydroxide + Water ⟶ Calcium hydroxide**

Metal oxides neutralise acids to form salt and water.

> **Calcium oxide + Hydrochloric acid ⟶ Calcium chloride + Water**

In the above reaction, calcium chloride is the salt.

Metal hydroxides of reactive metals can be produced by the reaction of metals and water.

> **Calcium + Water ⟶ Calcium hydroxide + Hydrogen**

Some metal oxides show properties of both acids and bases. Such oxides are called **amphoteric oxides**. Aluminium oxide is a good example.

All metal hydroxides have two ions: a metal ion and a hydroxide ion.

EXERCISES

A Complete the following word equations.

1 zinc + sulfuric acid _____ + _____

2 calcium + hydrochloric acid _____ + _____

3 iron + sulfuric acid _____ + _____

4 aluminium + hydrochloric acid _____ + _____

5 magnesium + sulfuric acid _____ + _____

B Write the name and the formula of the salt produced when:

1 Magnesium reacts with hydrochloric acid.

Name: _____ Formula: _____

2 Copper reacts with nitric acid.

Name: _____ Formula: _____

3 Zinc reacts with nitric acid.

Name: _____ Formula: _____

4 Aluminium reacts with sulfuric acid.

Name: _____ Formula: _____

ISBN: 978-0-17-018952-1

5 Aluminium reacts with hydrochloric acid.

Name: _____ Formula: _____

6 Calcium reacts with hydrochloric acid.

Name: _____ Formula: _____

7 Calcium reacts with sulfuric acid.

Name: _____ Formula: _____

C Give scientific reasons for the following.

1 It is dangerous to do experiments with lithium metal and hydrochloric acid.

_____

_____

_____

2 Hydrogen gas is no longer used to fill balloons.

_____

_____

_____

3 Hydrogen gas is collected by the downward displacement of water.

_____

_____

_____

D Write word equations and balanced chemical equations for the following.

1 Zinc metal reacts with dilute hydrochloric acid.

Word equation _____

Balanced chemical equation _____

2 Calcium metal reacts with dilute sulfuric acid.

Word equation _____

Balanced chemical equation _____

E Study the experiment below to answer the questions.

A student places a few drop of dilute hydrochloric acid onto magnesium ribbon.

Dilute HCl

Magnesium

Dish

1 State the observations that the student must record while doing this experiment.

_____

_____

_____

_____

_____

_____

_____

2 A chemical reaction occurs in the dish. Name the reactants and products of this chemical reaction.

Reactants _____ and _____

Products _____ and _____

ISBN: 978-0-17-018952-1

3 One of the products formed in this reaction is a salt. Describe a practical way to separate this salt from the solution.

_____

_____

_____

4 The salt formed is an ionic compound. State the symbols of the two ions that make up this compound.

_____ and _____

5 The other product formed is a gas. State what makes it difficult for the student to observe this gas.

_____

_____

_____

F Study these two experiments to answer the questions.

Experiment 1

Green copper compound

Limewater

Tube 1

A green copper compound is heated strongly. The gas formed during this reaction is bubbled through limewater. The limewater goes milky.

Experiment 2

Sulfuric acid

Green copper compound

Tube 2

Sulfuric acid is added to a green copper compound. A blue solution is formed in the test tube. The gas produced during this reaction is bubbled through limewater. The limewater goes milky.

1 Limewater in both tubes turns milky. Give an explanation for this colour change.

_____

_____

_____

_____

_____

2 State the name and the formula of the green copper compound used in these experiments.

_____

_____

3 The green copper compound in tube 1 becomes a black powder. Write down the name and formula of this black compound.

Name: _____ Formula: _____

4 Write a word equation and a balanced chemical equation for the reaction occurring in Tube 1.

_____

_____

_____

_____

ISBN: 978-0-17-018952-1

5 A blue solution forms in Tube 2. Write down the name and formula of this blue compound.

Name: _____ Formula: _____

6 Write a word equation and a balanced chemical equation for the reaction occurring in Tube 2.

_____

_____

_____

_____

G **A student heated a marble chip as shown below.**

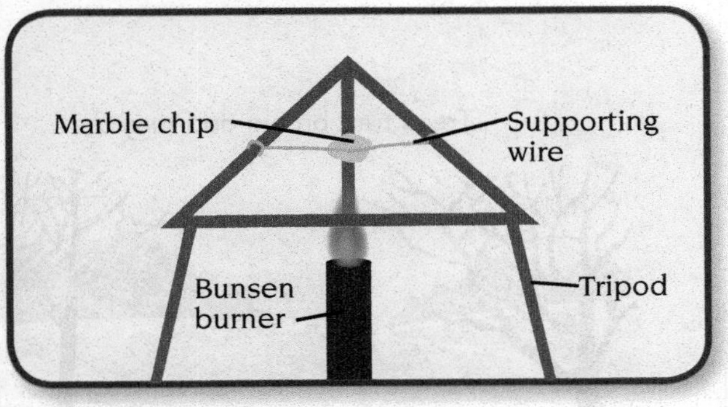

Marble chip

Supporting wire

Bunsen burner

Tripod

1 Write down the chemical name and formula of marble.

Name _____

Formula _____

2 After heating, the student observed a white powdery substance on the marble chip. Write the name and formula for this substance.

Name _____

Formula _____

3 The student collected some of the white powder formed in the reaction, then dissolved it in water. Write the name and formula for the new substance formed when this white powder reacts with water.

Name _____ Formula _____

4 What is the common name for the above solution?

_____

5 Heating marble produces a colourless gas. Name this gas.

_____

6 When dilute hydrochloric acid is added to marble chips, the same gas as in the above experiment is produced. Write a word equation and a balanced chemical equation for the reaction between marble chips and hydrochloric acid.

Word equation

_____

Balanced chemical equation

_____

7 The gas produced in the above reaction can be identified by using limewater. With the help of chemical equations explain how limewater reacts with this gas.

_____

_____

_____

_____

8 Baking soda contains sodium hydrogen carbonate and cream of tartar (powdered tartaric acid). Explain how baking soda helps the dough rise during baking.

_____

_____

_____

_____

ISBN: 978-0-17-018952-1

13 ACIDIC OXIDES AND ACID RAIN

Non-metal oxides are acidic in nature. They dissolve in water to form acids. Fossil fuels like coal, oil or petroleum products contain traces of sulfur as impurities. When fossil fuels burn in factories and in other internal combustion engines, they produce gases like **sulfur dioxide, sulfur trioxide** and **nitrogen dioxide**. These gases dissolve in water vapour in the rainclouds to form **sulfurous** or sulfuric acid and nitric acid. These acidic vapours then condense and fall as acid rain.

2 Oxides of sulfur and nitrogen are put into the air.

1 Fossil fuels burn in factories and internal combustion engines.

3 These oxides dissolve in water vapour.

4 Water vapour in clouds condense to fall as acid rain.

Acid rain destroys buildings.

It destroys paint work. It erodes marble sculptures and stone buildings. It speeds up metal corrosion.

Trees turn brown and may die.

Water in the soil, lakes and ponds become more acidic.

Nutrients in the soil are washed away or chemically altered.

Pond weeds, fishes and other aquatic life will be affected.

Acid rain is a problem mainly in industrial areas of Europe, USA, India and China. Wind can carry the acidic fumes from one area to another area. Fumes from USA are carried by wind to Canada where they cause acid rain.

Acidic water in a pond looks crystal clear because acid destroys algae and other aquatic organisms. Normal rain water is slightly acidic (pH 6). Acid rain can be as strong as lemon juice (pH 2).

Sulfuric acid is manufactured by burning sulfur, then dissolving the sulfur trioxide gas in water. Sulfuric acid is used in the manufacturing of **super-phosphate**, which is an important fertiliser.

ISBN: 978-0-17-018952-1

A Answer the following.

1 Name two gases responsible for acid rain.

_____

_____

2 State the main sources of these gases.

_____

_____

3 Normal rain can be slightly acidic. Suggest a reason for this.

_____

_____

4 Give scientific reasons for the following.

    a Acid rain causes more severe damage to limestone buildings than to concrete buildings.

_____

_____

    b Lakes and ponds, where acid rain falls, look crystal clear.

_____

_____

5 Write a paragraph summarising the major cause and the bad effects of acid rain.

_____

_____

_____

_____

_____

_____

_____

_____

ISBN: 978-0-17-018952-1

MECHANICS

SPECIFIC LEARNING OUTCOMES

✓ Calculate the speed and acceleration of an object in motion by using appropriate formulae.

✓ Draw and interpret distance-time and speed-time graphs from data obtained in motion experiments.

✓ Describe the motion of an object by studying distance-time graphs.

✓ Calculate the speed of an object from the gradient of a distance-time graph.

✓ Calculate the acceleration of an object from the gradient of a speed-time graph.

✓ Calculate the distance travelled by an object from its speed-time graph.

✓ Identify and describe the forces acting on a stationary object, an object moving at a constant speed, and an object accelerating or decelerating.

✓ Calculate the resultant force acting on an object by using appropriate formulae.

✓ Describe the relationship between force, mass and acceleration.

✓ Recognise weight as a force due to gravity and mass as a measure of the amount of matter.

✓ Describe the effect of friction on motion.

✓ Describe different forms of energy and identify energy transformations.

✓ Define pressure and explain the relationship between force, pressure and surface area.

✓ Calculate gravitational potential energy and kinetic energy by using appropriate formulae.

✓ Define work and power and calculate their values by using appropriate formulae.

ISBN: 978-0-17-018952-1

$$v_{average} = \frac{d}{t}$$

$$a = \frac{\text{change in speed}}{\text{change in time}} \quad \text{or} \quad a = \frac{\triangle v}{\triangle t}$$

$$E_p = mgh$$

$$F = ma$$

$$\text{Weight} = F_{gravity} = mg$$

$$E_k = \frac{1}{2}mv^2$$

$$\text{Work} = Fd$$

$$P = \frac{E}{t} \quad \text{or} \quad \frac{W}{t}$$

$$g = 10 \text{ ms}^{-2} \text{ (on Earth's surface)} \quad g = 10 \text{ Nkg}^{-1} \text{ (standard conditions)}$$

$$\text{Pressure} = \frac{F}{A}$$

Physics is the study of matter, energy, forces and their interactions. It involves the study of natural phenomena and events in quantities which can be measured. These physical quantities are measured in **units**. The units that we use to measure physical quantities are called the **SI** (System International) **units**.

Some of the commonly used physical units of measurement and their symbols are given in this table.

Physical quantity	Unit	Symbol
distance (d)	metre	m
time (t)	second	s
speed (velocity) (v)	metres per second	ms^{-1}
acceleration (a)	metre per second squared	ms^{-2}
mass (m)	kilogram	kg
energy (E)	joule	J
work (W)	joule	J
power (P)	watt	W
force (F) and weight (F_g)	newton	N
current (I)	ampere/amps	A
potential difference /voltage (V)	volt	V
electric resistance (R)	ohms	Ω
pressure	pascal	Pa

In some instances, when the quantity being measured is too big or too small, you may need to use bigger or smaller units of measurements. The most commonly used prefixes are:

Giga $= 10^9$ (billion)
Mega $= 10^6$ (million)
Kilo $= 10^3$ (thousand)
Centi $= 10^{-2}$ (hundredth)
Milli $= 10^{-3}$ (thousandth)
Micro $= 10^{-6}$ (millionth)

The Manapouri Power Station in Fiordland can produce 850 million watts of electricity. This is 850 megawatts (850MW).

ISBN: 978-0-17-018952-1

1 Complete the crossword puzzle below.

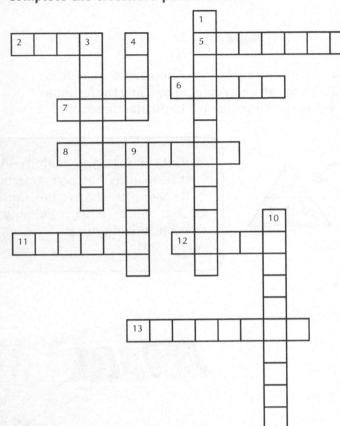

Across

2 One of the physical quantities measured in joules.

5 This quantity is measured in amperes (A).

6 The SI unit for measuring distance or length.

7 The SI unit for measuring potential difference.

8 This quantity is measured in pascals (Pa).

11 The SI unit for measuring weight and other forces.

12 The SI unit for measuring energy and work.

13 This physical quantity can be measured in metres per second.

Down

1 The symbol for the unit of measurement of this quantity is ms^{-2}.

3 The SI unit for measuring mass.

4 The SI unit for measuring power.

9 The SI unit for measuring time.

10 Ohm is the unit used for measuring this quantity.

2 State what physical quantity is being expressed in units described in each of the measurement below.

Example   2 L of milk _____ volume _____

a   10 kg of potatoes _____

b   A student walks 2 km to school _____

c   A car travels at 50 kmh^{-1} _____

d   Water boils at 100 °C _____

e   A 100 W light bulb is much brighter than a 60 W bulb _____

f   The food in the can gives you 1500 kJ _____

3 Express the following measurements in SI units.

a   10 minutes   _____   b   3 kilometres   _____

c   1500 g   _____   d   300 mm   _____

e   1.2 kilowatts   _____   f   250 mA   _____

4 A student calculated the following quantities correctly but failed to write the SI units. Complete his answers by writing the correct SI unit beside each answer.

a   The net force needed to push the van = 3500 _____

b   The height of the post = 2.5 _____

c   Gravitational potential energy gained by the ball = 850 _____

d   Weight of an astronaut on moon = 136 _____

e   Mass of the astronaut on moon = 80 _____

f   The power rating of the electric heater = 10 000 _____

ISBN: 978-0-17-018952-1

SPEED

When an object is moving we say it is in **motion**. When an object is not moving we say it is **stationary** or **at rest**. All moving objects have speed. Speed can be calculated by using the formula below. The symbol for speed is **v**, which stands for **velocity**, the speed in a particular direction.

Speed (v) = $\dfrac{\text{Distance travelled (d)}}{\text{Time taken (t)}}$

$v = \dfrac{d}{t}$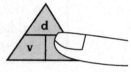

Velocity is a vector quantity. A vector quantity has two aspects – it has size and a direction. Speed is a scalar quantity. A scalar quantity has a size but no direction.

For rearranging this formula, try this 'triangle' method.

- Write down your formula in this triangle format.

- Use your finger to cover the quantity you want to find.

Write down your formula like this:

Time (t) = $\dfrac{d}{v}$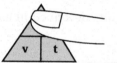

Distance (d) = v x t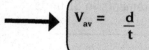

The **SI** unit for distance is metres (m) and time is seconds (s), so the unit for measuring speed is **metres per second.** This can be best written as **ms⁻¹**. If the distance is in **kilometres** and the time is in **hours**, the unit for speed can be written as **kilometres per hour**, or **kmh⁻¹**.

The speed of this van is 50 kmh⁻¹. This means, in one hour this van travels 50 kilometres. If it travels at a **constant** or steady speed it will travel 100 km in two hours. However, it is difficult for a vehicle to travel at a constant speed for long. In such situations we calculate the **average speed**. The formula below can be used to find the average speed.

Average speed = $\dfrac{\text{Total distance travelled}}{\text{Total time taken}}$ $\longrightarrow$ $V_{av} = \dfrac{d}{t}$

Example

Calculate the average speed of a rally car that travels 200 m in 80 seconds and another 80 m in 20 seconds.

Total distance travelled (d) = 200 m + 80 m = 280 m

Total time taken (t) = 80 s + 20 s = 100 s

Average speed (v_{av}) = $\dfrac{d}{t}$ = $\dfrac{280}{100}$ = **2.8 ms⁻¹**

ISBN: 978-0-17-018952-1

DISTANCE-TIME GRAPHS

If a car travels at a constant speed of 50 kmh⁻¹ for seven hours and you record the distance and time on a chart, you will get data like this.

Data like this can be used to plot **distance-time graphs**. Distance is usually marked on the vertical axis and time on the horizontal axis.

If you use the data to draw a distance-time graph, the shape of the graph will look like this.

Time (h)	Distance (km)
0	0
1	50
2	100
3	150
4	200
5	250
6	300
7	350

Distance-time graph

The distance-time graph for an object moving at a **constant speed** is always a straight line with a slope.

Shapes of distance-time graphs

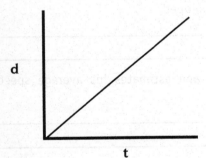

Distance-time graph for an object moving at a steady or constant speed.

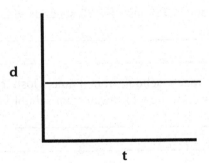

Distance-time graph for a stationary object.

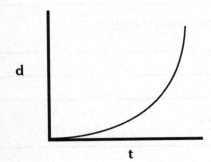

Distance-time graph for an accelerating object.

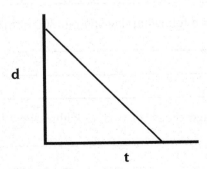

Distance-time graph for an object that is returning (at a constant speed) to its starting point.

ISBN: 978-0-17-018952-1

A Complete the table below by calculating the missing quantities.

Distance (m)	Time (s)	Speed (ms⁻¹)
500	25	
81		9
	7	50
55		11
250	20	

B Solve the following problems. (You must show all steps and the final units.)

1 A student walks 90 m to school and it takes 180 s. Calculate how fast the student walked.

_____

_____

2 A car travels at an average speed of 80 kmh⁻¹ for 3 hours. Calculate how far it has travelled.

_____

_____

3 A horse gallops at 8.5 ms⁻¹ for 20 seconds. Calculate how far it has galloped.

_____

_____

4 In order to reach school from home, Josh cycles for 10 minutes and estimates his average speed as 2.5 ms⁻¹. Calculate how far away from school he lives.

_____

_____

5 a At school athletics, Leah ran 250 m in 50 s. What was her average speed?

_____

_____

_____

b If on the following day she ran 300 m in 30 s. By how much did her speed increase?

_____

_____

_____

6 If the average speed a car is 45 kmhr⁻¹, how far can it travel in 40 min?

_____

_____

_____

ISBN: 978-0-17-018952-1

C Below is a distance-time graph for a car journey.

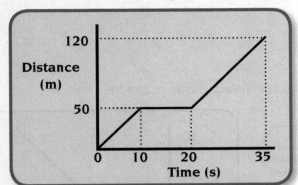

1 Describe the motion of the car between 0 and 10 seconds.

_____

_____

2 Describe the motion of the car between 10 and 20 seconds.

_____

_____

D To fill in time on the family's holiday trip, Kate noted the distances. Every 30 seconds for 4 minutes she recorded the distance the car had travelled. Below is the data she collected.

Time (s)	0	30	60	90	120	150	180	210	240
Distance (km)	0.0	0.1	0.3	0.6	1.0	1.5	?	2.5	3.0

1 In the space provided, draw a line graph using the above data. Label the axes first.

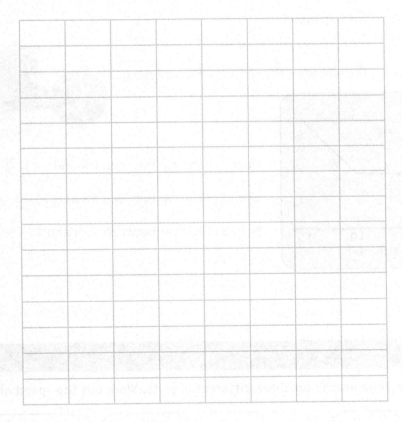

2 State what would be the best title for this graph.

_____

3 Use this graph to determine how far the car had travelled in 180 s.

_____

_____

4 Describe the motion of the car between

a 0 – 120s _____

b 120s – 240s _____

ISBN: 978-0-17-018952-1

The **gradient** or **slope** of a distance-time graph gives you the speed. Slope or gradient can be calculated by using the formula below.

$$\text{Gradient} = \frac{\text{Increase in vertical height (of the line)}}{\text{Increase in horizontal length (of the line)}}$$

Example 1

The graph on the right is a distance-time graph for the motion of a radio-controlled model car. The shape of the graph tells that the car is travelling at a constant speed. You can also calculate the speed of the car at various time intervals by finding the slope or gradient of this line graph. For example, the speed of the car between 0s and 2s is the gradient of the line AB.

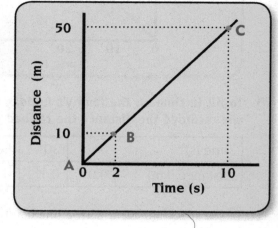

Gradient of the line $\text{AB} = \frac{10 \text{ m}}{2 \text{ s}} = \textbf{5 ms}^{-1}$

Example 2

Below is a distance-time graph showing the motion of a model car.

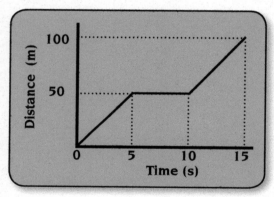

Speed of the car between 0 s and 5 s $= \frac{50 \text{ m}}{5 \text{ s}} = \textbf{10 ms}^{-1}$

Speed of the car between 5 s and 10 s $= \frac{0 \text{ m}}{5 \text{ s}} = \textbf{0 ms}^{-1}$

Speed of the car between 10 s and 15 s $= \frac{50 \text{ m}}{5 \text{ s}} = \textbf{10 ms}^{-1}$

EXERCISES

A **Below are distance-time graphs for three different objects. Work out the speed of each object.**

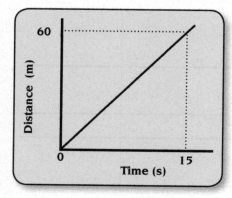

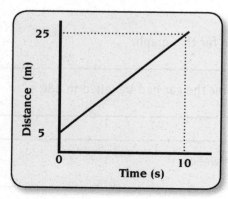

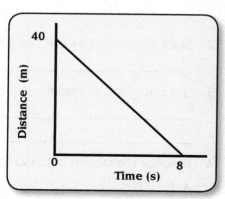

Speed = _____

Speed = _____

Speed = _____

ISBN: 978-0-17-018952-1

B The distance-time graph below represents the motion of a car.

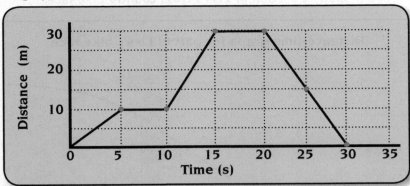

1 Calculate the speed of the car

   a between 0 and 5 seconds. _____ **b** between 5 and 10 seconds. _____

   c between 10 and 15 seconds. _____ **d** between 15 and 20 seconds. _____

   e between 20 and 30 seconds. _____

   f Calculate the average speed of the car for the whole journey. _____

2 Complete the chart below by describing the motion of the car during the time interval given. The first row has been done for you.

TIME INTERVAL (S)	DESCRIPTION OF MOTION
0–5	The car is travelling at a constant speed of 2 ms^{-1}
5–10	
10–15	
15–20	
20–30	

C The graph below shows the distance-time graphs of two cars X and Y, travelling on the same road.

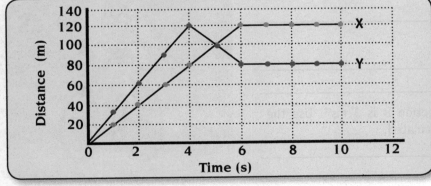

1 Describe the journey of car X.

_____

2 Describe the journey of car Y.

_____

3 State at what time the cars pass each other. _____

4 State at what time both cars stopped during the race. _____

5 State how far apart the cars were when they stopped. _____

6 Calculate the speed of car X for the first 5 seconds.

_____

_____

ISBN: 978-0-17-018952-1

D The distance-time graph below shows a section of Lisa's cross country bike race.

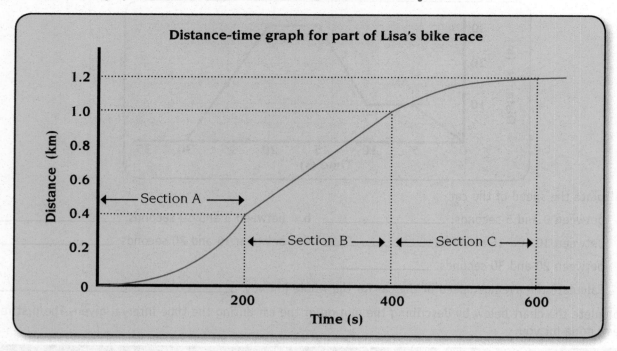

Distance-time graph for part of Lisa's bike race

a How far does Lisa travel in the first 400 seconds? State the answer in metres.

_____

_____

b Calculate Lisa's average speed over the whole 600 s. State the answer in ms^{-1}.

_____

_____

c Describe the motion of Lisa and her bike during section A and B.

Section A:

_____

_____

Section B:

_____

_____

d Lisa's coach says her speed during Section B is 3 ms^{-1}. Use the graph above to show how it can be calculated.

_____

_____

_____

_____

ISBN: 978-0-17-018952-1

ACCELERATION AND SPEED-TIME GRAPHS

ACCELERATION

When an object increases its speed, we say the object is **accelerating**. When an object slows down or decreases its speed, we say the object is **decelerating**. A decelerating object has negative acceleration.

Acceleration can be calculated by using the formula below.

$$\text{Acceleration} = \frac{\text{Change in speed}}{\text{Change in time}}$$

$$\text{OR} \quad a = \frac{\Delta v}{\Delta t}$$

Δv means change in speed = final speed – initial speed
Δt means change in time.

We could also write the formula for acceleration as follows.

$$\text{Acceleration} = \frac{\text{Final speed - Initial speed}}{\text{Time taken}}$$

If the speed is in **ms⁻¹** and the time is in **s**, the unit for acceleration will be metre per second per second = ms⁻¹s⁻¹. This can be best written as **ms⁻²**.

Example 1
A racing car speeds up from 8 ms⁻¹ to 24 ms⁻¹ and takes 8 s to do so. Calculate its acceleration.

Final speed   = 24 ms⁻¹
Initial speed   = 8 ms⁻¹
Change in speed = 24–8 = 16 ms⁻¹
Time   = 8 s
Acceleration $= \frac{16\text{ms}^{-1}}{8\text{s}} = \mathbf{2\ ms^{-2}}$

SPEED-TIME GRAPHS

If you can measure the speed of a moving object at regular time intervals, a speed-time graph for that object can be drawn.

A speed-time graph for a car ride can be drawn by taking the speedometer reading at a regular time interval. The data below shows the results of such an experiment.

A line graph (time on the horizontal axis and speed on the vertical axis) drawn for this data gives you the speed-time graph.

TIME (s)	SPEED (ms⁻¹)
0	0
30	10
60	20
90	30
120	40
150	50
180	50
210	50
240	50
270	50

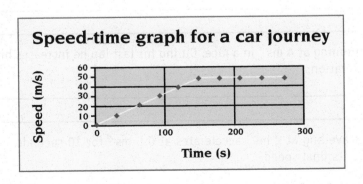

Speed-time graph for a car journey

The above graph shows that the car is speeding up or accelerating from start until it reaches the speed of 50 ms⁻¹. The car then keeps the same speed (constant or steady speed) for the rest of its journey.

ISBN 978 0 17 018952 1

Shapes of speed-time graphs

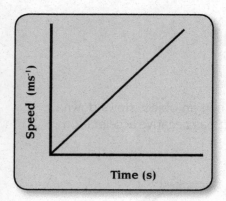

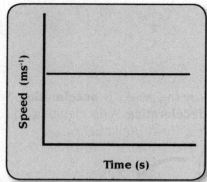

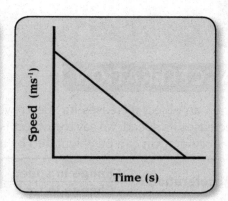

| A graph of this shape tells that the object is accelerating (constant acceleration). | A graph of this shape tells that the object is having a constant speed or no acceleration. | A graph of this shape tells that the object is decelerating constantly (negative acceleration). |

EXERCISES

A Use appropriate formulae to calculate the unknown quantities.

1 A racing car is travelling at 10ms⁻¹. Ten seconds later it is travelling at 25 ms⁻¹. Calculate the acceleration of the car. (Show all workings and units.)

_____

_____

2 A sports car accelerates at 4 ms⁻² starting from rest. It then reaches a speed of 8 ms⁻¹. Calculate how long it takes to reach this speed.

_____

_____

3 A car is travelling at 2 ms⁻¹. Five seconds later it is travelling at 12 ms⁻¹. Calculate its acceleration.

_____

_____

4 A rocket lifts off from a launch pad from rest and reaches a speed of 500 ms⁻¹ in 10 seconds. Find its acceleration.

_____

_____

5 A motor bike speeds up from 5 ms⁻¹ to 13 ms⁻¹ and takes 4 s to do so. Calculate its acceleration.

_____

_____

6 Tyler is running at 4 ms⁻¹ in a race. During his last lap he increases his speed to 5 ms⁻¹ in 2 seconds. Calculate his acceleration.

_____

_____

7 A truck travelling at 2 ms⁻¹ accelerates at 0.5 ms⁻² for 10 seconds. Calculate its final speed.

_____

_____

ISBN: 978-0-17-018952-1

B Below is a speed-time graph for a racing car's trial run.

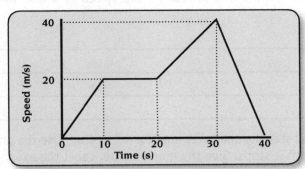

1 Describe the journey of the car during the time intervals given in the chart below.

Time interval (s)	Description of motion
0–10	
10–20	
20–30	
30–40	

2 What was the highest speed achieved by the car? _____

3 What was the speed of the car at 15 s? _____

4 What was the acceleration of the car at 15 s? _____

5 For how long did the car travel at a steady speed? _____

6 How long did the whole journey last? _____

C Study the speed-time graph below to answer the questions.

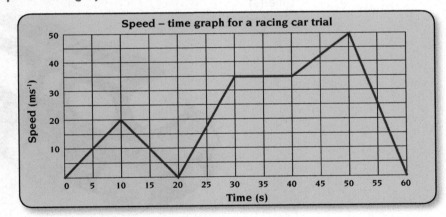

1 Write T (true) or F (false) beside each statement about the car's motion.

  a The car was accelerating between 0 and 10 s.

  b Between 10 s and 20 s the car was not moving.

  c Between 20 s and 30 s the car was accelerating.

  d Between 30 s and 40 s the car was not moving.

  e Between 40 s and 50 s the car was accelerating.

  f After 50 s the car went backwards.

  g The whole journey took 60 s.

  h The acceleration between 30 s and 40 s was zero.

  i The highest speed was reached at 50 s.

ISBN: 978-0-17-018952-1

D Alicia's brother has a toy car which moves along a vinyl floor when he winds up the spring and lets the car go. The car accelerates constantly from rest to reach a top speed of 4 ms⁻¹ in 8 seconds.

1 Calculate the car's acceleration.

_____

_____

_____

_____

2 From its top speed, the toy car slows steadily and takes another 4 s to stop. Use the grid below to sketch the toy car's motion from its release until it stops. Give the graph a title. Label the axes.

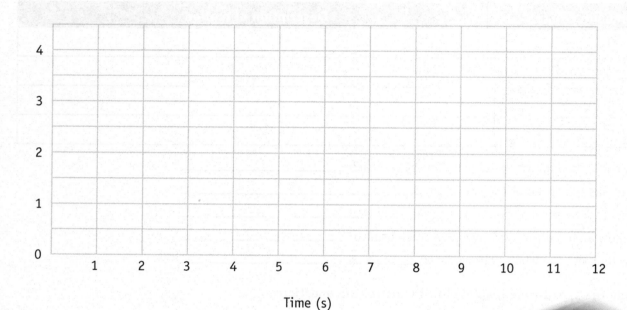

ISBN: 978-0-17-018952-1

GRADIENT OF SPEED-TIME GRAPHS

The gradient or slope of a speed-time graph gives you the acceleration. Slope or gradient can be calculated by using the formula below.

$$\text{Gradient} = \frac{\text{Increase in vertical height (of the line)}}{\text{Increase in the horizontal length (of the line)}}$$

Example 1

The graph on your right is a speed-time graph for the motion of a racing car. By looking at the shape of the graph, you can tell that the car is accelerating. You can also calculate the acceleration at various times.

For example, the acceleration between 0 and 2 s is the gradient of the line AB.

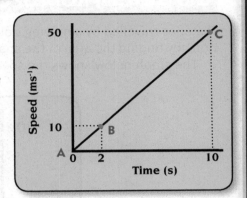

Gradient of the line AB = $\dfrac{10\ \text{ms}^{-1}}{2\ \text{s}}$ = **5 ms^{-2}**

Acceleration of the bike between 2 and 10 s is the gradient of the line BC.

Gradient of the line BC = $\dfrac{50-10}{10-2} = \dfrac{40\ \text{ms}^{-1}}{8\ \text{s}}$ = **5 ms^{-2}**

TICKER-TIMERS

A ticker-timer can be used to study short and fast motion of small objects such as a trolley or a toy car. When you switch on the ticker-timer it makes a certain number of dots in a certain period of time. This is usually marked on the timer as its **frequency**. The frequency of most ticker-timers is 50 Hz. This means these ticker-timers can make 50 dots in one second.

The diagram below shows how ticker-timers work.

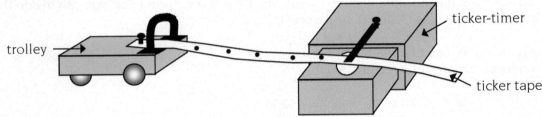

Below is a middle section of ticker-tape that recorded the motion of a trolley.

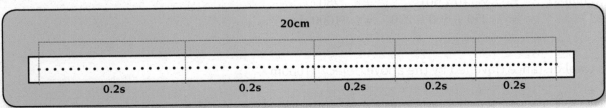

This ticker-timer makes 50 dots in one second. Therefore, it makes 10 dots in 0.2 s.
Total length of the tape (distance) = 20 cm = 0.20 m (This can be measured by using a ruler.)
Total time shown is 1 s, although speed changes.

The average speed of the trolley over the whole second $= \dfrac{d}{t}$ $= \dfrac{0.2\ \text{m}}{1\ \text{s}}$ **= 0.2 ms^{-1}**

ISBN: 978-0-17-018952-1

The ticker-tape obtained can also be used for describing the motion of the object.

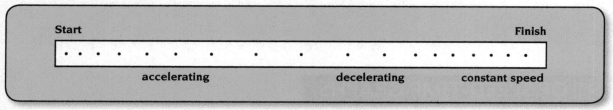

- Evenly spaced dots means constant speed.
- If the space between dots is widening progressively, then the object is accelerating.
- If the space between dots is decreasing progressively, then the object is slowing down or decelerating.

DISTANCE FROM SPEED-TIME GRAPHS

Distance travelled by a moving object can be calculated from its speed-time graph of motion. This is done by finding the area of the shape/s beneath the graph.

The graph below shows the speed-time graph of a moving object.

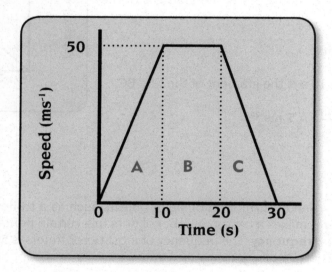

For calculating the total distance travelled by this object in 30 seconds, you need to calculate the total area beneath the line. Reason: distance = speed × time. For convenience, you can divide this into three areas, two triangles (A and C) and a rectangle (B).

Formula for finding the area of a triangle = ½ bh
Formula for finding the area of a rectangle = lb

Area of triangle A	=	½ (10 x 50)	=	250
Area of rectangle B	=	10 x 50	=	500
Area of triangle C	=	½ (10 x 50)	=	250
Total area	=	250 + 500 + 250	=	1000

Therefore total distance travelled by the above object =1000m

ISBN: 978-0-17-018952-1

A **The table below shows the speed of two racing cars recorded during a pre-race trial.**

1 Plot speed-time graphs for their journeys using the data supplied. You may use two different colours to draw two lines on the same axes. Give the graph a title and label both axes.

Time (s)	Speed (ms^{-1})	
	Car 1	Car 2
0	0	0
1	5	10
2	10	20
3	15	30
4	20	40
5	25	50
6	30	40
7	35	30
8	40	20
9	40	10
10	40	0

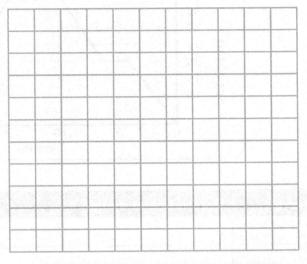

2 Describe the motion of car 1.

3 Describe the motion of car 2.

4 Calculate the acceleration of car 1 for the first 8 seconds.

5 Calculate the acceleration of car 1 between 8 and 10 seconds.

6 Calculate the acceleration of car 2 between 5 and 10 seconds.

B **Given below are four ticker-tapes showing the motion of four different trolleys. Describe the motion of each trolley.**

Start   Finish

1 ·· · ·   · · · · ··

2 · · · · · · · · · · ·

3 · · · · · · ·   ·   ·   ·

4 · · · · · · · · ·

C Study the speed-time graph given below to complete the chart.

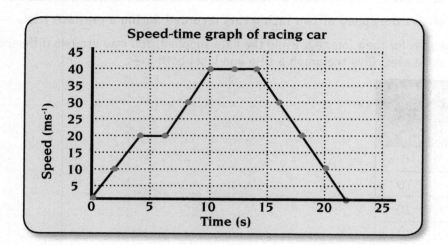

Speed-time graph of racing car

Time interval	Description of motion	Acceleration
0–4 s		
10–14 s		
14–22 s		

D An old ute develops a small leak in its fuel tank as it drives along an unsealed country road. The diagram below shows the pattern of the diesel drops on the road. Later the driver found out that the ute was losing a drop every 3 seconds.

START

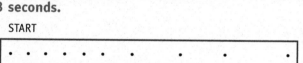

1 On the axes below, sketch (shape only) a distance-time graph and a speed-time graph for this part of the truck's journey.

Distance

Time

Speed

Time

2 During the journey, the ute changes its speed from 2 ms⁻¹ to 8 ms⁻¹ at one particular time interval (between two diesel drops). Calculate the acceleration of the ute.

_____

_____

_____

_____

ISBN: 978-0-17-018952-1

E The data below show the speed and time measurement of a soccer player during a soccer game.

Time (s)	Speed (ms⁻¹)
0.0	0.0
0.5	1.5
1.0	3.0
1.5	4.5
2.0	6.0
2.5	6.0
3.0	6.0
3.5	6.0
4.0	3.0
4.5	0

1 Plot a speed-time graph for the motion of the soccer player during the game. Label the axes. The speed axis should show 0, 1, 2, 3, 4, 5, 6 ms⁻¹.

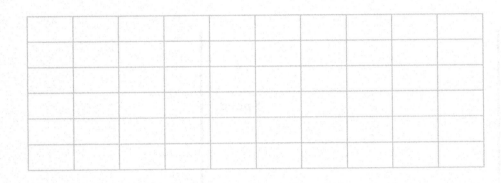

2 State the maximum speed of the soccer player during his run.

_____

3 Describe his motion between 2.0 s and 3.5 s.

_____

_____

_____

4 Use the graph to calculate the player's acceleration between 0 s and 2.0 s.

_____

_____

_____

_____

5 Use the graph to calculate the total distance travelled by the player during this run.

_____

_____

_____

_____

_____

ISBN: 978-0-17-018952-1

F The ticker-tape given below shows the motion of a shark.

START   FINISH

Study this and then draw (shape only) a distance-time graph and a speed-time graph.

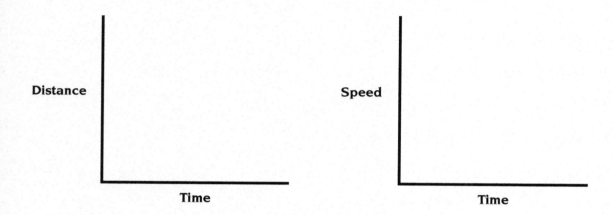

Distance   Speed

Time   Time

G Below is a speed-time graph for Cody on a training run. Calculate the total distance he ran in 60 seconds.

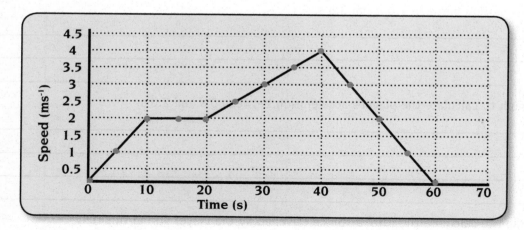

Calculate the total distance covered in 60 seconds _____

_____

_____

_____

_____

ISBN: 978-0-17-018952-1

FORCE

A force is a push or a pull or a twist. Forces have both **magnitude** and direction. The SI unit for measuring force is the **newton** (**N**). There are **contact** forces and **non-contact** forces. When you push a box along the floor, there is contact. When a magnet attracts and pulls an iron nail towards it, there is no contact at first.

Contact force

Non-contact force

Forces can do the following things:

- A force can stop a moving object.
- A force can accelerate or decelerate a moving object.
- A force can change the direction of motion of a moving object.
- A force can change the shape of an object.
- A force can keep an object in place.
- A stationary object starts moving when a force acts on it.

If a force pushes an object to the right and another force of the same magnitude pushes it to the left, these two forces will cancel each other out and we say the forces are **balanced**. The net force or the **resultant** force will be zero and the object will remain **stationary**. When balanced forces act on a moving object, the object will move at a constant speed in a straight line.

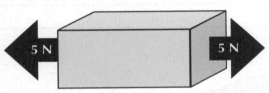

Resultant force = 5 – 5 = 0 N

Balanced forces

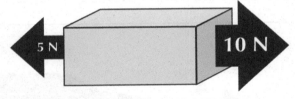

Resultant force = 10 – 5 = 5 N to the right

Unbalanced forces

The diagram below shows the action and reaction forces acting on a car moving at a steady speed.

Friction between the surface of the road and the tyres, and air resistance (drag force).

The driving force of the engine.

Support force = the upward push of the road.

Gravity or gravitational force. This is also called the weight force.

The above example shows that forces work in pairs. These pairs can be called **action and reaction forces**.

EXERCISES

A Complete the chart below.

Forces acting on object	Net force acting on object	Direction of movement
14 N ← ☐ → 6 N		
 5 N ← ☐ → 5 N		
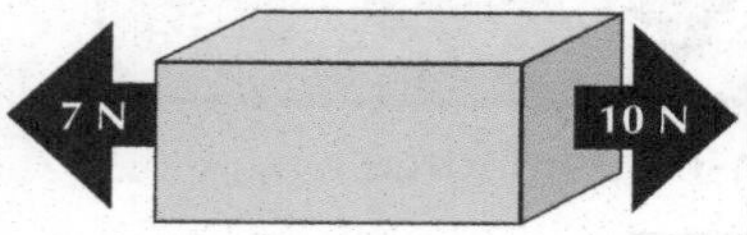 7 N ← ☐ → 10 N		
 12 N ← ☐ → 10 N		

B The diagram below shows two major forces, A and B, acting on a stretch limo that moves at a constant speed in a forward direction in a straight line for five minutes during its journey.

1 The magnitude for the force A during this particular section of its journey is 5000N. What would be the magnitude of force B during this section of its journey? What causes force B?

_____

_____

2 Describe what would happen to the limo if force A becomes greater than force B for another 5 minutes of its journey.

_____

_____

3 Describe what would happen to the limo if force B becomes greater than force A.

_____

_____

ISBN: 978-0-17-018952-1

C Complete the following sentences.

1 A force can be a push, _____, or a twist.

2 When an object remains stationary, all the forces acting on it are _____.

3 The SI unit for measuring force is _____.

D This soccer ball begins slowing down the instant it leaves the boot.

1 The forces acting on this ball are _____.
   balanced/unbalanced

2 When a player kicks the ball, this action may change the

   _____ and the speed of the ball.

E Sam bikes to work every day. The four main forces acting on his bike are: weight, friction, support and push.

1 On the diagram above draw arrows to show the direction of all four forces acting on Sam and his bike. Identify and label each of these forces.

2 At one point on his ride, the four forces combine to give a net force of zero. Explain why Sam and his bike continue moving at a constant speed even though this net force is zero.

_____

_____

_____

_____

_____

ISBN: 978-0-17-018952-1

The relationship between force, mass and acceleration is summarised in the formula below.

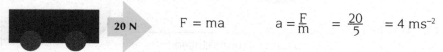

> **Force = mass x acceleration**
> **F = ma**

Here, **F** = force in newtons (N), **m** = mass in kilograms (kg) and **a** = acceleration in ms^{-2}

This is a mathematical way of stating that 'more force increases acceleration' and 'more mass reduces acceleration'.

In the example below, a trolley of mass 5 kg is pulled with a force of 20 N.

20 N $F = ma$ $a = \dfrac{F}{m} = \dfrac{20}{5} = 4 \text{ ms}^{-2}$

A large force will produce greater acceleration.

When the mass is doubled, the acceleration is halved. Two trolleys of mass 5 kg each are tied together and pulled with a force of 20 N. In this example, the force remains the same but the mass doubles.

5 kg
5 kg   **20 N** $F = ma$ $a = \dfrac{F}{m} = \dfrac{20 \text{ N}}{10 \text{ kg}} = 2 \text{ ms}^{-2}$

This F = ma relationship between force, mass and acceleration is known as Newton's Second Law of Motion.

EXERCISES

A   Use the appropriate formula to work out unknown quantities.

1   An unbalanced force of 40 N acts on a model boat of mass 8 kg. Calculate the acceleration of the boat.

_____

_____

2   Calculate the force needed to accelerate a model truck of mass 5 kg at 4 ms^{-2}.

_____

_____

3   An acceleration of 5 ms^{-2} was produced on a trolley by a force of 45 N. Calculate the mass of the trolley.

_____

_____

4   A car of mass 1500 kg speeds up from 10 ms^{-1} to 14 ms^{-1} in 8 s.

   a   Calculate the acceleration of the car. _____

   b   Calculate the unbalanced force that is causing the acceleration. _____

_____

5   A trolley of mass 5 kg moves at a steady speed in a straight line. A force meter attached to it reads 20 N. State the size of the reaction force acting on this trolley.

_____

_____

ISBN: 978-0-17-018952-1

B The photograph shows a cricket ball after being hit by a player. Choose a word from the key list to complete each statement about the moving cricket ball.

KEY LIST: volume, mass, speed, acceleration, friction, force

1 _____ of 0.16 kg.

2 _____ of 200 N.

3 _____ causes the ball to slow.

4 _____ of 3 ms⁻¹.

C Complete the chart below.

Force (f)	Mass (m)	Acceleration (a)
	25 kg	4 ms⁻²
50 N	2 kg	
1000 N		25 ms⁻²
	5000 g	2.8 ms⁻²
440 N		8 ms⁻²
	500 g	10 ms⁻²
180 N	90 kg	

D Sina was investigating the relationship between force and acceleration. She used the ticker-timer and the trolley as shown in the diagram below. Different pulling forces were used to accelerate the trolley.

Ticker-timer

Trolley

Pulling force

Results

Trial 1

Trial 2

Sina then drew a speed-time graph for trial 1 after studying the ticker-tape she obtained.

1 Sketch the results for trial 2 on the same axes provided.

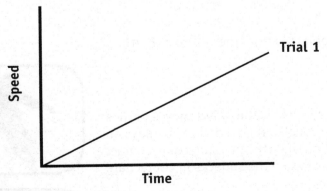

Trial 1

Speed

Time

2 Describe what the slope of the speed–time graph represents.

_____

_____

3 Write a conclusion for Sina's experiment.

_____

_____

ISBN: 978-0-17-018952-1

MASS, WEIGHT AND GRAVITY

In everyday life, we sometimes use the words 'mass' and 'weight' as though they are the same. In reality, mass and weight are not the same.

Mass is a measure of the total amount of matter in an object. The SI unit for mass is kilogram (kg). Mass does not change from place to place.

Weight is the **force** (pull) of gravity on a mass. The SI unit is **newton** (**N**). Weight can vary from place to place, depending on the force of gravity.

Gravity causes any free-falling object to **accelerate** downwards. On Earth this **acceleration** is about 9.8 ms^{-2}. We call this acceleration 'g', and usually round it to 10.

$$g = 10\,ms^{-2}$$

For weights, Newton's Second Law of Motion, F=ma can be written as:

$$F_{gravity} = mg$$

On Earth, the weight of 1kg mass can be easily calculated in newtons.

$$F_{gravity} = mg$$
$$= 1\,kg \times 10\,ms^{-2}$$
$$= 10\,N$$

The illustration below shows the weight of a 50 kg mass measured on Earth, in space and on the moon. The mass of the object remains the same but the weight changes, due to the difference in gravity.

Mass = 50 kgs
g = 10 ms^{-2}
Weight = 50 x 10 = 500 N

Mass = 50 kgs
g = 0 ms^{-2}
Weight = 50 x 0 = 0 N

Mass = 50 kgs
g = 1.7 ms^{-2}
Weight = 50 x 1.7 = 85 N

The gravitational pull on an object depends on two factors:
- the mass of the object
- the distance or how far the object is from the centre of Earth.

FRICTION

Friction is a contact force that occurs when one object moves or tries to move over another object or through a fluid. Friction is always an **opposing** force. There is less friction when the two surfaces in contact are smooth. Friction will be greater when the surfaces in contact are rough.

Very little friction

Sometimes friction can be a nuisance but in many situations you cannot survive without friction. Some disadvantages of friction are:
- Friction opposes motion.
- Friction causes wear and tear in moving machine parts.
- Friction produces heat energy.

Great friction

ISBN: 978-0-17-018952-1

Friction can be reduced either by **lubricating** the surfaces or by using ball-bearings or roller-bearings. Grease and oil are common lubricants.

When a skydiver falls, gravitational force and frictional force act at the same time but in opposite directions. When these two forces balance each other and before the parachute opens, the skydiver will reach maximum speed, called **terminal velocity**.

EXERCISES

A The table below shows the value of g (acceleration due to gravity) at different places. Use this information to answer questions 1–7. (All quantities are measured in ms⁻² units.)

Place	Earth	Space	Jupiter	Mars	Mercury	Venus
g	10	0	26	4	0.3	9

The photo on the right shows a satellite. It has a mass of 200 kg on earth.

1 Calculate the **weight** of this satellite on earth.

_____

2 State the **mass** of this satellite in space.

_____

200 kg

3 Calculate the **weight** of this satellite in remote space.

_____

4 On which planet would this satellite weigh the most?

_____

5 On which planet would this satellite weigh the least?

_____

6 What would the weight of this satellite be on Mars?

_____

7 On which planet would the least force be needed to lift this satellite?

_____

B **Study the following situations, then complete the sentences by selecting appropriate words.**

1 **Parachuting**

Friction is _____ (an advantage/a nuisance) here because friction between the balloon and the air _____ (slows down/speeds up) the downward motion for safe landing.

ISBN: 978-0-17-018952-1

2 Space shuttle

Friction is _____ (an advantage/a nuisance) here because friction between the outer surface of the space shuttle and the air generates _____ (kinetic/heat) energy.

C Complete the table below. (The first row has been completed for you.)

Situation	Friction an advantage or nuisance	Reason
Walking	Advantage	Stops us from slipping
Ice skating		
Braking a car		
Rock climbing		
Swimming		

D Match up the words in list A with meanings or explanations in list B.

List A
1 newtons
2 kilograms
3 gravity
4 weight
5 mass
6 balanced force
7 unbalanced force
8 contact force
9 non-contact force
10 friction

List B
A A measure of the amount of matter in an object.
B A reaction force which slows down a rolling ball.
C SI unit for measuring force and weight.
D SI unit for measuring mass.
E A type of force which can act at a distance.
F Pull of large masses like Earth and other planets.
G A type of force where a contact is required.
H Keeps an object at rest, or moving at a steady speed.
I Changes the speed or direction of a moving object.
J A force due to the pull of gravity.

Answers

1	2	3	4	5	6	7	8	9	10

E Write T (true) or F (false) beside each sentence

1 The mass of an object will be the same as its weight.

2 The SI unit for measuring weight is Newtons.

3 The mass of an astronaut will be the same on Earth and in space.

4 Friction always produces heat energy.

ISBN: 978-0-17-018952-1

9 FORCES AND PRESSURE

Pressure is the force per unit area applied in a direction perpendicular to the surface of an object.

$$\text{Pressure} = \frac{\text{Force}}{\text{Area}}$$

$$P = \frac{F}{A}$$

Force (F) is measured in newtons (N) and the surface area is measured in m^2 so the SI unit for pressure is **Nm^{-2}**. This is also called the **pascal (Pa)**.

The relationship between pressure and the surface area can be explained by the illustration below.

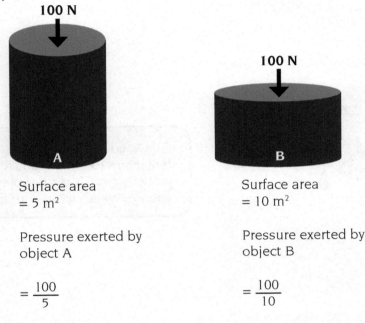

100 N

100 N

A

B

Surface area
= $5\ m^2$

Surface area
= $10\ m^2$

Pressure exerted by object A

Pressure exerted by object B

$= \dfrac{100}{5}$

$= \dfrac{100}{10}$

$= 20\ Nm^{-2}$

$= 10\ Nm^{-2}$

The pressure of a given force decreases as the surface area increases. In other words, reducing the area increases the pressure.

EXAMPLE

A block of polystyrene with dimensions 6 m x 5 m x 10 m has a mass of 600 kg. Calculate the maximum pressure exerted by this block when placed on a flat surface as shown below. (g = 10 Nkg^{-1}).

Weight force ($F_{gravity}$)   = mg
                        = 600 x 10 = 6000 N

Surface area of contact = 6 x 5 = 30 m^2

$P = \dfrac{F}{A}$

$= \dfrac{6000}{30}$

$= 200\ Nm^{-2}$

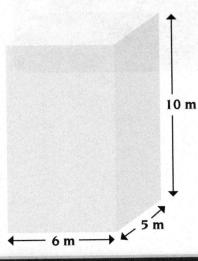

10 m

5 m

6 m

ISBN: 978-0-17-018952-1

The narrow metal blade under this ice skating boot exerts a very high pressure on the surface of ice. This pressure causes the ice to melt under the blade. This allows the skater to glide smoothly.

Skis used on a mountain slope have a larger surface area than ice skates. This larger area reduces the pressure on the snow preventing the skier from sinking.

The wide tyres of a tractor reduce the pressure exerted on soft ground, so they do not sink into the ground.

ISBN: 978-0-17-018952-1

A   Two identical blocks have a mass of 80 kg each, and dimensions of 2 m x 4 m x 8 m. They are placed side-by-side on a plasticene surface in two different ways, A and B.

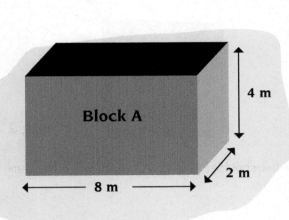

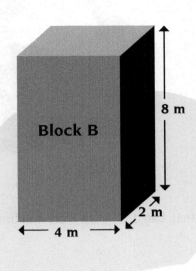

1   Calculate the weight of the blocks in newtons.

_____

_____

2   Calculate the pressure exerted by Block A.

_____

_____

_____

_____

3   Calculate the pressure exerted by Block B.

_____

_____

_____

_____

4   Which block will leave a deeper mark in the plasticene? Explain.

_____

_____

_____

_____

_____

_____

ISBN: 978-0-17-018952-1

B Bag 1 has a wide shoulder strap while Bag 2 has a narrow shoulder strap.

Bag 1

Bag 2

Holly finds that Bag 1 is much more comfortable on her shoulder than Bag 2, even when the same mass is put inside.

1 Explain the reason why Bag 1 is more comfortable than Bag 2. Your explanation must contain the words: **force**, **pressure** and **surface area**.

_____

_____

_____

_____

2 Delete the incorrect word in this sentence:

> The pressure of a given force decreases / increases as the surface area decreases.

C Two kinds of boot are shown below. Boot B has metal studs on its sole.

Boot B

Boot A

1 Explain why the design of Boot B provides better grip on the ground than Boot A.

_____

_____

_____

_____

_____

ISBN: 978-0-17-018952-1

D The diagram below shows part of one of the front tyre tread patterns of a truck parked in the goods delivery bay of a supermarket.

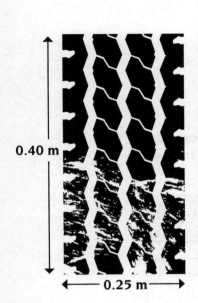

0.40 m

0.25 m

Because of the treads, only approximately 50% of the tyre's area is in contact with the ground. The force exerted by this tyre is 8088 N over the area drawn here. Calculate the pressure exerted by this tyre on the ground. Show all steps in your working. State the units for your final answer.

_____
_____
_____
_____
_____
_____
_____
_____
_____
_____
_____
_____
_____
_____

ISBN: 978-0-17-018952-1

Energy is not easy to define, but we can state the following:
- Energy comes in different forms (see below)
- Energy can be changed from one form to another
- Energy is not a substance – although substances like petrol contain energy.

FORMS OF ENERGY

Energy exists in different forms. Two major types of energy are **potential energy** and **kinetic energy**. Potential energy is stored energy. Energy present in food and fuels are examples of potential energy. All moving objects have kinetic energy.

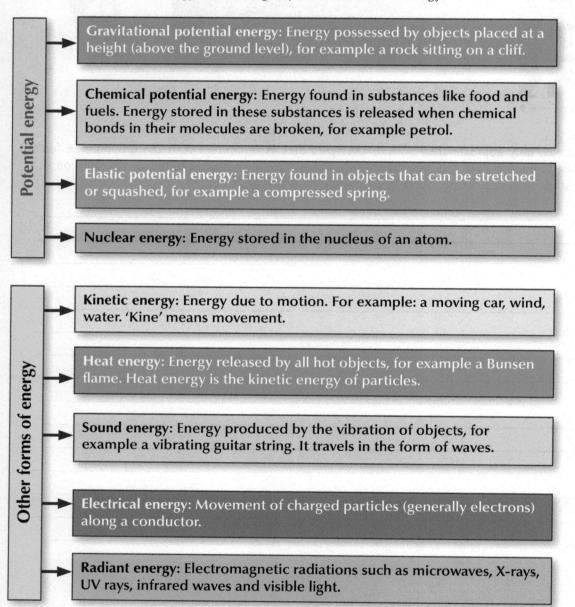

Potential energy

Gravitational potential energy: Energy possessed by objects placed at a height (above the ground level), for example a rock sitting on a cliff.

Chemical potential energy: Energy found in substances like food and fuels. Energy stored in these substances is released when chemical bonds in their molecules are broken, for example petrol.

Elastic potential energy: Energy found in objects that can be stretched or squashed, for example a compressed spring.

Nuclear energy: Energy stored in the nucleus of an atom.

Other forms of energy

Kinetic energy: Energy due to motion. For example: a moving car, wind, water. 'Kine' means movement.

Heat energy: Energy released by all hot objects, for example a Bunsen flame. Heat energy is the kinetic energy of particles.

Sound energy: Energy produced by the vibration of objects, for example a vibrating guitar string. It travels in the form of waves.

Electrical energy: Movement of charged particles (generally electrons) along a conductor.

Radiant energy: Electromagnetic radiations such as microwaves, X-rays, UV rays, infrared waves and visible light.

ENERGY TRANSFORMATIONS

The law of conservation of energy states that energy cannot be created or destroyed, but that any form of energy can be transformed into almost any other form.

ISBN: 978-0-17-018952-1

Study the photographs below.

Radiant energy (solar/light)

Chemical energy

Chemical energy

Kinetic energy

Nuclear reactions in the Sun produce radiant energy. Plants use this energy to make food and store it as chemical potential energy. Animals that eat plants store this chemical energy in their muscles and fat. We use meat and plants as food. Food is our source of physical energy for all activities. When we move around, this stored chemical potential energy transforms mainly into kinetic energy.

Energy transformation can be summarised in the form of an 'energy equation' as shown below.

Solar (light) energy ⟶ Chemical potential energy ⟶ Kinetic energy

During the energy transformation, some energy is always lost in the form of heat energy. Machines are energy converters but no machine is 100% energy-efficient.

EXERCISES

A   Complete the following sentences by using the words from the list below.

List: kinetic, potential, substance, work, forms, food, radiant

1   Energy is not a _____.
2   Energy exists in different _____.
3   Energy is the ability to do _____.
4   All living things obtain their energy from the _____ they eat.
5   Green plants use the _____ energy from the sun to make food.
6   All moving objects have _____ energy.
7   Stored energy is called _____ energy.

B   Write down the forms of energy present in the regions marked A, B and C.

Water particles in the beaker have

_____ energy.

The flame of this Bunsen burner has

_____ energy.

Gas used in this burner has

_____ energy.

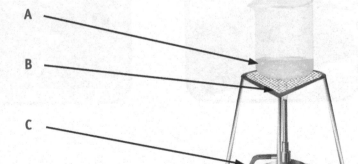

A
B
C

C   Write energy equations for the following situations.

1   A girl switches on her MP3 player and listens to her favourite music.

_____ ⟶ _____ ⟶ _____

ISBN: 978-0-17-018952-1

2 An athlete picks up a shotput from the ground, holds it above his shoulder, then throws it.

_____ ⟶ _____ ⟶ _____

3 Aran ran up a sand-dune as part of his fitness training.

_____ ⟶ _____ ⟶ _____

D **Match up the forms of energy in list A with explanations or examples given in list B.**

List A
1   Chemical potential energy
2   Gravitational potential energy
3   Nuclear energy
4   Radiant energy
5   Kinetic energy
6   Heat energy
7   Solar energy
8   Electrical energy
9   Elastic potential energy
10 Sound energy

List B
A   Energy present in a compressed spring.
B   Energy found in a live electric cable.
C   Vibrating objects make this form of energy.
D   Radiant energy from the Sun.
E   Energy found in a bar of chocolate.
F   A book sitting on a shelf has this form of energy.
G   Found in the nucleus of a uranium atom.
H   A flying frisbee has this form of energy.
I   X-rays and microwaves are examples of this type of energy.
J   An electric stove makes this form of energy.

Answers

1	2	3	4	5	6	7	8	9	10

E   **Write down the major forms of energy possessed or given off by the following objects or situations.**

1 _____   2 _____   3 _____

4 _____   5 _____   6 _____

ISBN: 978-0-17-018952-1

F Complete the table below.

Energy input	Energy converter	Main energy output
Electrical energy	1	Heat energy
2	Lawnmower engine	3
Chemical potential energy	4	Sound energy
5	Solar water-heater	6
7	Windmill	8
Electrical energy	9	Kinetic energy
10	Catapult	11
12	Nuclear power station	13

G Match each energy converter with the correct energy conversion for which it is designed.

Energy converter

1 Bow and arrow
2 Flashlight
3 Spotlight
4 Electric jug
5 Television set
6 A car
7 Nuclear power station
8 Steam engine
9 Water stored in a dam
10 A microphone
11 A loudspeaker

Energy conversion

A Electrical energy into sound energy.
B Nuclear energy into heat energy.
C Elastic potential energy into kinetic energy.
D Chemical potential energy into light energy.
E Sound energy into electrical energy.
F Electrical energy into light energy.
G Electrical energy into heat energy.
H Electrical energy into light and sound.
I Chemical potential energy into kinetic energy.
J Gravitational potential energy into kinetic energy.
K Heat energy into kinetic energy.

Answers

1	2	3	4	5	6	7	8	9	10	11

H For each energy equation, write down the name of an object designed for such a transformation.

1 Kinetic energy ⟶ sound energy _____

2 Electrical energy ⟶ kinetic energy _____

3 Electrical energy ⟶ heat energy _____

4 Chemical potential energy ⟶ electrical energy _____

5 Electrical energy ⟶ sound energy _____

6 Elastic potential energy ⟶ kinetic energy _____

ISBN: 978-0-17-018952-1

KINETIC ENERGY

Every object in motion has kinetic energy. When a moving object stops suddenly, its kinetic energy transforms mainly into heat energy. When you apply brakes on a fast-moving car, the brake discs get hot.

Energy of all kinds is usually measured in **joules (J)**. One Joule is a very small amount of energy. Kinetic energy can be calculated by using the formula below.

$$E_k = \tfrac{1}{2}\, mv^2$$

E_k = kinetic energy in joules (J), **m** = mass in kilograms (kg), **v** = speed in ms^{-1}.

Study the following to see what happens to kinetic energy when the speed of the object doubles.

Mass of the car = 1000 kg	Mass of the car = 1000 kg
Speed of the car = 5 ms^{-1}	Speed of the car = 10 ms^{-1}
E_k = $\tfrac{1}{2}$ mv^2	E_k = $\tfrac{1}{2}$ mv^2
= $\tfrac{1}{2}$ × 1000 kg × 5 ms^{-1} × 5 ms^{-1}	= $\tfrac{1}{2}$ × 1000 kg × 10 ms^{-1} x 10 ms^{-1}
= 12,500 J	= 50,000 J
= 12.5 kJ	= 50 kJ

In the above example the mass of the car remains the same. When the speed doubles, the kinetic energy increases by a factor of four.

Example

Calculate the kinetic energy of a tennis ball of mass 0.15 kg travelling at a speed of 10 ms^{-1}.

E_k = $\tfrac{1}{2}$ mv^2
m = 0.15 kg
v = 5 ms^{-1}
E_k = $\tfrac{1}{2}$ × 0.15 kg × 10 ms^{-1} × 10 ms^{-1} = 7.5 J

GRAVITATIONAL POTENTIAL ENERGY

When you lift a box to the top shelf, the box gains gravitational potential energy. Gravitational potential energy can easily be changed into kinetic energy. When the box falls from the shelf, its gravitational potential energy transforms into kinetic energy, heat and sound when it hits the floor.

Gravitational potential energy can be calculated by using the formula below.

$$E_p = m\,g\,h$$

Here **m** = mass in kilograms, **g** = acceleration due to gravity, which is 10 ms^{-2}, and **h** = height in metres.

The athlete at this position has more gravitational potential energy than kinetic energy.

ISBN: 978-0-17-018952-1

We have already seen that m × g = weight force in newtons, so we can write the formula for gravitational potential energy as shown below.

> **gravitational potential energy = weight force x height**

Study this example.
The gravitational potential energy of a rugby ball of mass 400 g at 12 m height above ground can be calculated as follows.

Mass of the ball $= 400$ grams
$\qquad\qquad\qquad = 0.4$ kg
$\mathbf{g}$ on earth $= 10$ ms^{-2}

Height above ground $= 12$ m
$\qquad$ ball $E_p = m\,g\,h$
$\qquad\qquad\quad = 0.4$ kg $\times 10$ ms$^{-2} \times 12$ m
$\qquad\qquad\quad = 48$ J

EXERCISES

A Complete the table below.

Object	Mass (kg)	Velocity (ms^{-1})	Kinetic energy (J)
1	1.25	2	
2	900	5	

B The diagram below shows a radio-controlled model car of mass 0.8 kg. The car travels from A to D. From A to B the car travels at a constant speed of 2 ms^{-1}. From B to C the car travels at a constant speed of 4 ms^{-1}. The car stops at D.

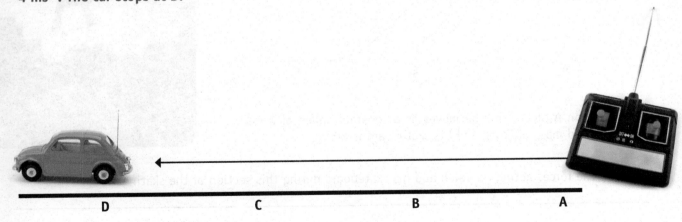

D   C   B   A

1 Calculate the kinetic energy of the car when it was between A and B.
_____

2 Calculate the kinetic energy of the car when it was between B and C.
_____

3 Explain what happened to the kinetic energy of the car when its speed doubled.
_____

4 Explain what happened to the kinetic energy of the car when it stopped at point D.
_____

ISBN: 978-0-17-018952-1

C A car travels at a constant speed of 5 ms⁻¹ from A to B. Then it climbs a small hill. The car reaches the top of the hill and stops at point C for a while. The car has a mass of 800 kg.

1 Name the major type of energy the car has at

    a   point B. _____

    b   point C. _____

2 Calculate the weight of the car.

_____

3 Calculate the amount of kinetic energy the car has at point B.

_____

4 If the amount of potential energy the car has at C is the same as the kinetic energy at B, calculate the height of the hill from the ground level.

_____

_____

D **The diagram below shows the profile of a skate rink that Caleb uses on weekends.**

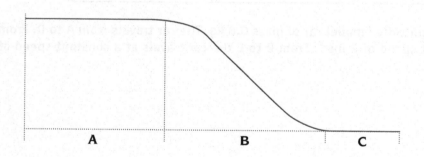

Along section A of the rink he moves at a constant speed of 2 ms⁻¹.
The combined mass of Caleb and his skateboard is 80 kg.

a Describe the forces acting on Caleb and his skateboard during this section of the skating.

_____

_____

b Calculate the kinetic energy of Caleb and his skateboard in section A.

_____

_____

c He then skates down section B of the rink with a constant acceleration of 1.2 ms⁻². Calculate the net force acting on Caleb and his skateboard during this section.

_____

_____

ISBN: 978-0-17-018952-1

E The diagram below shows a pendulum of mass 0.5 kg swinging from A to C. Study the information given in the diagram to answer the following.

1 Name the major type of energy the pendulum has at point A.

_____

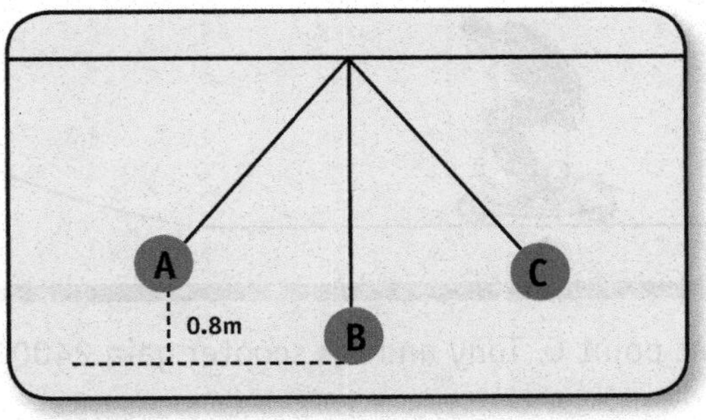

2 Calculate the amount of energy the pendulum has at point A (compared to B).

_____

_____

_____

3 Describe the energy transformations that occur as it swings from A to C.

_____

_____

F A tool box of mass 20 kg is lifted up from the ground onto a ute.
The height of the lift from the ground is 1.5 m.

1 Calculate the weight of the tool box

_____

2 Calculate the gravitational potential energy gained by the tool box when it is placed on the ute.

_____

_____

_____

G An empty ski-lift has a mass of 50 kg. Troy and Briar travelled in a ski-lift between points A and B. Troy and Briar are 65 kg and 55 kg respectively. Point A is 2 m above ground level and point B is 8 m above ground level.

1 State the total mass of the ski-lift and the two passengers.

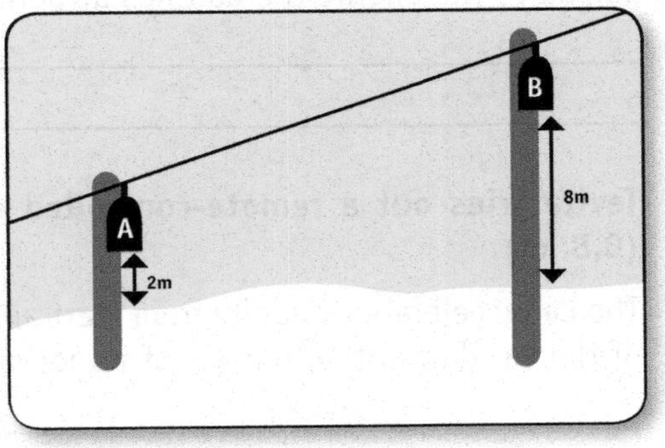

_____

2 Calculate the gravitational potential energy of the ski-lift and its two passengers at point A.

_____

_____

3 Calculate the gravitational potential energy of the ski-lift and its two passengers at point B.

_____

4 Use answers from questions **2** and **3** to explain the difference in the amount of gravitational potential energy gained by the ski-lift at point A and point B. (Note: the heights of 2m and 8m are measured from the same level.)

_____

_____

_____

_____

_____

ISBN: 978-0-17-018952-1

H Tony and his scooter have a total mass of 80 kg. He rides his scooter as shown in the diagram below.

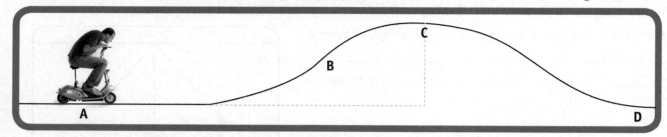

1 At point C, Tony and his scooter gain 2400 J of energy. Calculate how high point C is from ground level.

_____

_____

2 As he rides downhill from C to D he applies the brakes and the scooter stops at point D. State the major energy transformation that occurs during this part of the journey.

_____

_____

I Study the diagram, then complete the following sentences.

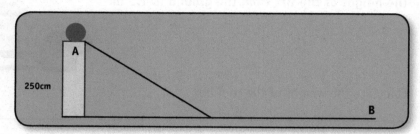

A steel ball of mass 5 kg is rolled from a height of 250 cm. The ball rolls down the ramp and stops at point B.

1 The steel ball has a weight of _____ N. (How much?)

2 At point A the ball has _____ energy. (What type?)

3 In between points A and B the ball has more _____ energy. (What type?)

4 Calculate the energy the ball has at point A.

_____

_____

J Tevita tries out a remote-controlled car on a concrete driveway. The toy car has a mass of 800 g (0.8 kg).

1 The car accelerates steadily from start and gains a maximum speed of 2 ms^{-1} in 4 s. Calculate the acceleration of the car during the first 4 s of its journey.

_____

_____

2 Calculate the size of net force acting on the car during its first 4 s journey.

_____

_____

ISBN: 978-0-17-018952-1

The car continued its journey for another 4 s with the same speed in a straight line and then it decelerates steadily and stops in 2 seconds as Tevita steers the car into a gravelled area.

3 Use this information and the information from question 1 to draw a speed-time graph for the 10 s journey of the radio-controlled car.

4 Between 4 s and 8 s the toy car's speed remains constant. Explain what the constant speed tell us about the forces acting on the car.

_____

_____

_____

_____

_____

_____

_____

_____

5 Discuss why the car decelerates when it is driven into the gravelled area. Consider driving force and friction, and how they affect the net force.

_____

_____

_____

_____

_____

_____

_____

_____

_____

_____

ISBN: 978-0-17-018952-1

WORK

When you push a supermarket trolley you are doing physical work. When you do work, you transfer energy from one place to another, either in the same form or in a different form. Work is done whenever an energy transfer occurs.

Work can be calculated by using the formula below.

> **Work = force x distance (in the direction of the force)**

> **W = f x d**

W	
f	d

Here **f** is force in newtons (N) and **d** is distance in metres (m). The unit for work is joules (J).

Lifting these books from the ground involves physical work in overcoming weight force.

Sweeping leaves is also physical work.

Example

A boy of mass 50 kg climbs the stairs as shown in the diagram. Calculate the amount of work he does.

$W = f \times d$
$\quad = m \times g \times h$
$\quad = 50 \text{ kg} \times 10 \text{ ms}^{-2} \times 2 \text{ m}$
$\quad = 1000 \text{ J}$

2m

POWER

Power can be defined as the rate of doing work. Power is the rate at which energy is transferred per second. Power can be calculated by using the formula below.

> **Power = $\dfrac{\text{Work}}{\text{Time}}$**

> **P = $\dfrac{W}{t}$**

W	
P	t

The SI unit for power is watts (W). Work is in joules (J) and time is in seconds (s) 1 watt = 1 joule per second, work done is equal to energy gained by an object. So you can also use the formula below to work out power.

> **P = $\dfrac{E}{t}$**

E	
P	t

E = energy

ISBN: 978-0-17-018952-1

Study the example below.

This crane lifts a concrete slab of mass 500 kg to a height of 5 m. It takes 20 s to complete this task. The power (work rate) of this crane can be calculated by following the steps below.

$$\begin{aligned} \text{Power} &= \frac{\text{Work}}{\text{Time}} \\ &= \frac{mgh}{t} \\ &= \frac{500 \text{ kg} \times 10 \text{ Nms}^{-2} \times 5 \text{ m}}{20 \text{ s}} \\ &= 1250 \text{ W} \end{aligned}$$

EXERCISES

A Complete the table below by writing either 'Work done' or 'No work done'.

	Action	Work done/No work done
1	Picking up a box from the ground.	
2	Turning the pages of a book.	
3	Pushing against a tree, which does not move.	

B A professional weightlifter is lifting a mass of 100 kg. During his first trial he lifts the mass to a height of 1.2 m.

1 Calculate the weight of the mass (in newtons) he is lifting.

_____

2 Calculate the amount of energy gained by the mass at this position.

_____

3 Calculate the amount of work done by the weightlifter.

_____

4 Name the type of energy the iron mass has at this position (see diagram).

_____

5 During his second trial he lifts the same mass to a height of 2 m. How much more extra work does he have to do in this activity?

_____

_____

_____

ISBN: 978-0-17-018952-1

C Dylan and Jack go for a scooter ride. Dylan has a mass of 60 kg and Jack has a mass of 50 kg. The scooter they ride on has a mass of 130 kg. They travel up a slope as shown in the diagram below.

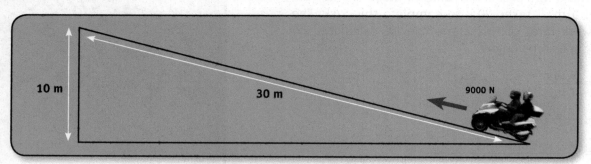

1 State the total mass of the scooter and its passengers.

_____

_____

_____

2 Calculate the total amount of energy gained by the scooter and two people at the top of the slope.

_____

_____

_____

_____

3 The force needed to push the scooter up the slope is measured as 9000 N. Calculate the work done by the scooter and its passengers as it travels up the slope.

_____

_____

_____

_____

4 The energy gained by the scooter and two people and the amount of work done are not equal. Discuss why there is a difference in these two values.

Include the following points in your answer.

• The type of energy the scooter and its passengers gain at the top of the slope.

• The value of both energy and work done and the reason why there is a difference.

_____

_____

_____

_____

_____

_____

_____

_____

ISBN: 978-0-17-018952-1

D Lisa is about to climb the stairs as shown in the diagram. Her mass is 50 kg and she is carrying two bags. Each bag has a mass of 5 kg.

5m

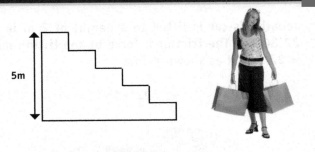

1 Calculate the total weight of Lisa and her bags.

_____

2 How much work does Lisa have to do to reach the top of the stairs?

_____

_____

E Gerome has a mass of 50 kg. He runs up stairs at school in 9 seconds. His friend Tim also has a mass of 50 kg. He runs up the stairs in 8 seconds.

1 State who you think does the most work. _____

2 State who you think has more power. Give a reason for your answer.

_____

_____

F Complete the puzzle below to find out the definition for power.

| 1 | 2 | 3 |

| 4 | 5 | 1 | 3 |

| 5 | 1 |

| 6 | 2 | 7 | 8 | 2 |

| 6 | 9 | 4 | 10 |

| 7 | 11 |

| 12 | 9 | 13 | 3 |

1 The SI unit for force.

2 The SI unit for energy and work.

3 Amount of matter in an object.

4 The SI unit for distance.

5 An opposing force that stops a moving object.

6 The pulling force of great masses like Earth and other planets.

7 ms^{-2} is a unit for measuring this quantity.

8 ms^{-1} is a unit for measuring this quantity.

9 The SI unit for power.

10 All moving objects have this kind of energy.

11 Food and fuels contain this kind of energy.

12 This form of energy is wasted when friction occurs.

13 Another word for change.

kinetic energy – K

transformation – N

joules – H

gravity – W

heat – D

friction – A

acceleration – I

watts – O

newton – T

chemical energy – S

metres – R

mass – E

speed – C

ISBN: 978-0-17-018952-1

G Jonathan's car is lifted to a height of 2 m in a garage. The combined weight of car-lift and car is 22 800 N. The frictional force of the lifting mechanism is 80 N and the lifting force of the car-lift is 22 960 N as shown below.

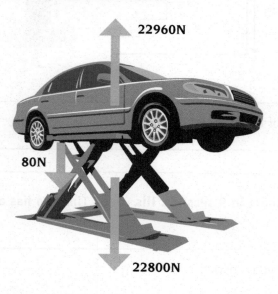

22960N

80N

22800N

1 Calculate the initial resultant force acting on the car and the car-lift.

_____
_____
_____
_____

2 State the combined mass of the car and the car-lift.

_____
_____
_____
_____
_____

3 Explain the difference between the terms 'mass' and 'weight' of the car.

_____
_____
_____
_____

4 The car is hoisted to a height of 2 m by the car-lift. Calculate the work done by the car-lift.

_____
_____
_____

H Two elevators in a hotel do similar jobs. Both elevators lift 12 passengers to the third floor in 10 seconds, but the power outputs of the two elevator motors are different. Explain how this can be so.

_____
_____
_____
_____
_____
_____
_____
_____
_____
_____
_____
_____

ISBN: 978-0-17-018952-1